IEE HISTORY OF TECHNOLOGY SERIES 21

Series Editor: Dr B. Bowers

EXHIBITING
ELECTRICITY

Other volumes in this series:

EXHIBITING
ELECTRICITY

—◆✳◆—

by

K. G. Beauchamp

THE INSTITUTION OF ELECTRICAL ENGINEERS

Published by: The Institution of Electrical Engineers, London, United Kingdom

© 1997: The Institution of Electrical Engineers

British Library Cataloguing in Publication Data

A CIP catalogue record for this book
is available from the British Library

ISBN 0 85296 895 7

Printed in England by Bookcraft, Bath

Contents

13 The modern era **275**

14 Trade fairs **299**

Preface

Throughout the 19th and 20th centuries technical exhibitions, held for the benefit of both cognoscente and the general public alike, have presented a mirror to the progress of science, engineering and increasingly, towards the second half of the 19th century and throughout the 20th century, to electrical technology.

Here the history of such public exhibitions is traced from their beginnings towards the end of the 18th century to the present day, with particular reference to their presentation of electrical invention and manufacture relative to the known state of knowledge at the time. The key factors determining this progression are described together with the influence of competing nations on the changing format of these presentations.

Contributions to this history will be considered broadly as a chronological sequence, modified where necessary to conform to a thematic chapter grouping of the exhibitions. The term, 'exhibition' will be used generically to indicate a fair, exposition, exhibition or a show. In general, expositions, exhibitions and world fairs are held for prestige or education, a show or trade fair for business and a public fair for entertainment, but all contain common elements.

We can identify three aspects of any exhibition, exposition or world fair of this kind as: displaying the skill and ingenuity of a nation; informing and educating the visitor; and entertainment or commercial influence, to which we might add in recent times the opportunity to express nationalism or to provide propaganda relative to a particular theme or culture. The 'mix' of these features has varied with the time and purpose of the exhibition and will be apparent from the content and description throughout the following Chapters.

Exhibitions themselves are important not only to provide a given generation with a summary of its current capability but also to present a forum for new inventions and techniques, often shown to the public for the first time. The special requirements of exhibitions are also productive of new developments which find use in other activities: sophisticated lighting techniques, many examples of which are discussed in the text, the use of dioramas, new concepts in architecture, the first use of moving pavements, exploitation of new ideas in mass travel, such as monorail or magnetic levitation railways, and finally new methods of crowd control, particularly in recent expositions where the attendance may have been in tens of millions during the period of the event.

The writer is grateful for a great deal of assistance received from very many individuals and organisations during the preparation of this book. Thanks are due to the following for providing much of the research material used: the British Library, Science Museum Library, Bodleian Library, Guildhall Library, Public Records Office, the Libraries of the Institutions of Mechanical and Electrical Engineers, Manchester and Lancaster City Libraries, and the Library of Lancaster University. Special thanks are due to Susan Bennett, the Library Administrator of the Royal Society of Arts, Clare Colvin, the archivist of the Royal Television Society, Larry Tise of the Franklin Institute, Leonore Symons, the archivist of the Institution of Electrical Engineers, Peter Briggs of the British Association, David Dacam of the Institution of Incorporated Executive Engineers, Henry Skinner of the British Radio and Electronic Manufacturer's Association, Bill Price of Electrex Ltd, Everil Robertson of World's Fair Inc, Monika Wiss of Interkama, the organisers of Expo'98 at Lisbon, Expo '2000 at Hannover, the BIE in Paris, John Perratory and David Watson of Calgary Expo'2005, Carl Malumud of Internet Multicasting Service, Jonathan Coopersmith of Texas A & M University, Jack Howlett of Oxford for French translations, Andrew Beauchamp for Internet access, Gordon Woodward of the IEE History of Technology Professional Group S7 for valuable information on Liverpool and its electrical exhibitions, and to Brian Bowers of the Science Museum for much helpful advice.

It has not proved possible to locate the present copyright holders of some of the earlier illustrations used, and sincere apologies are offered for the noninclusion of their names.

Ken Beauchamp

1996   Lancaster

Chapter 1
Origins

1.1 The exhibition

The tremendous changes in society activated by the Industrial Revolution brought with them the need to explain, educate and interest the people in the new artifacts made possible by invention, improved design and production techniques in ways not previously considered or even thought necessary. The reasons for this were twofold—the need to find markets for the newly manufactured goods and also to ensure a ready supply of skilled workmen to produce them.

Foremost in this process was 'the exhibition' in which the results of the ingenuity, design, skill and manufacture could be gathered together from a wide geographical area (initially national and later international) and shown to the public for their entertainment and education. With the third constituent, that of a showplace for a nation's skills and products, these three elements of the exhibition—entertainment, education, and proprietorial display—can be identified in all modern exhibitions, expositions and world fairs since the Great Exhibition of 1851. We can perhaps add a fourth notion, applicable to international exhibitions in recent times—that of political propaganda. The promulgation of a theme or political idea has become inseparable from the modern exposition and it is a matter of argument whether this adds to or detracts from its presentation.

The origins of the exhibition can be traced back directly to the needs of trade between peoples and nations in the middle ages. But the modern idea of the exhibition has only an indirect relationship with trade. We do not attend an exposition or world fair with the idea of purchasing any of the exhibits put before us. This idea of a fair without direct trade involvement could be considered as a new concept but, if we look into the development of the medieval fair, we find this is not without precedent in the centuries before industrial manufacture became widespread.

1.2 Trade in medieval Europe

Trade within and between nations had long been the province of itinerant merchants, sometimes travelling long distances with their trading goods to find a customer willing to barter. For convenience in the exchange of goods, and later credit, periodic gathering of merchants at specific locations began to take place early in medieval Europe. Since the recurring dates when these exchanges took place were often linked to religious festivals, they became known as 'fairs' (Latin: 'feriae' or feast) [1]. Fairs solved the early problems of distribution of goods for barter between these travelling traders and also provided the opportunity for the demonstration of skills and crafts and for the exchange of ideas.

They became a prominent economic institution of Europe in the middle ages and it is in this period that their number and diversity flourished. Early accounts of these fairs are sparse. One of these, from the 7th century, describes the yearly fair held in October for four weeks at the Abbey of St. Denis near Paris to honour the saint's day. It was opened by a procession of monks from the Abbey whilst the Parliament of Paris proclaimed a holiday, called Landi, in order that its members might '*take part in the great marriage feast of commerce and religion*'. The monks administered the event and received due levies on the transactions conducted there [2]. Although noted as a centre for trade in the local wine, honey, grain, wood, salt and dyes, the fair also became a meeting place for merchants from England, Spain, Saxony and other European locations bringing with them '*iron and lead of the Saxons, slaves of the northern nations, jewellary and perfume of the Jews, honey and madder from Austria and merchandise of Egypt and the East*' [3]. We can recognise in these 'commodity fairs' the market place of today with its multiplicity of trading stalls or perhaps the 'car boot sale' of very recent times with its coming together of itinerant traders.

1.3 Trade fairs

As the economic revival of Western Europe developed in the 10th and 11th centuries fairs increased in number, sited strategically along the overland routes connecting states, usually at a major crossroads, within cities and at some maritime ports. Of these the most important were the fairs established under the Counts of Champagne, held at various locations in France and arranged to follow each other in a cycle throughout the year. They were basically international and attracted merchants from all quarters of Europe, the Middle East and Africa. Each fair lasted about six weeks. The amount of trade carried out justified the establishment of money-changing booths, credit transactions and the settlement of merchants' quarrels. Credit facilities were introduced by Italian merchants in the 12th century in the form of bills of exchange or 'letters of fair'. Later it became commonplace to settle debts at the Champagne fairs, using the location as a clearing house for bills. It is also to these fairs that we owe many of our agreed weights and measures systems and regulations. One example is the troy weight for gold and silver, first established at the October fair of Troyes. Other fairs were in operation during the peak period for the fair in the 12th and 13th centuries. In Italy fairs were held at Pisa, Venice, Genoa, Padua and Milan, in Germany at Cologne, Hamburg, Leipzig, Frankfurt and Magdeburg, in Flanders at Bruges, Ypres and Lille, and in France, in addition to the Champagne fairs, at Rouen, Beauvais, Arras, Paris, Nimes, Montpellier and Lyon.

Particularly large fairs were held in Russia at Berdicheff, Elizavetgrad, Karkoff, Kursk, Riga and Rostof. The biggest was Nijni-Novgorod, an international fair, which was founded in 1366 and even after the conquest of the town by Ivan the Terrible in 1552,

'... *Monks of the monastery very cleverly made Nijni a place of religious as well as a commercial resort and levied taxes on the trade which they fostered*' [3].

After 1751 the tolls were collected by the state. The trade it managed was staggering. In the year 1790 it collected 36 million roubles, equivalent to £4 500 000 in equivalent currency of the day, with a daily throughput of visitors of 40 000.

Generally the fairs held in England were national in character and included entertainment, so providing a carnival atmosphere. The rights of foreign merchants in England were granted in the Magna Carta (1215) and confirmed by Henry III in 1216 as enabling the stranger to:

- come to England
- depart thereout
- remain

- travel by land and water
- buy and sell
- be free of evil tolls
- enjoy the ancient customs.

Stourbridge Fair, held on the river banks of Cambridge, is probably the oldest known English fair, founded on a charter granted by King John in 1211. The University of Cambridge had much jurisdiction over this fair and issued a proclamation just before its opening, known as *'crying the fair'* in which those taking part must obey certain instructions in the name of the King, namely:

1. *To make no 'fraye, cry, shrekinge or any noyse ... vexinge and disquietinge of ye King's leage people ... under pain of imprisonment and further punishment as offense shall require'*
2. *Scolars and Scolar's servants wear no weapons and strangers leave their weapons in their inn.*
3. *... various charges and commands to those supplying bread, beer, salt, fish, meat, etc. to avoid adulteration.*

The fair at Stourbridge became very extensive. Daniel Defoe, writing in 1723, claimed that [4],

> '... This Fair is not only the greatest in the whole Nation but in the World; nor, if I may believe those who have seen them all, is the Fair at Leipzig in Saxony, the Mart in Frankfurt on Main, or the Fairs of Nuremberg or Augsburg, may in any way compare to this Fair in Stourbridge'.

Other principal sites in England were at London, Chester, Boston, St.Ives and Bury St.Edmunds.

1.4 Sample fairs

At the beginning of the 14th century, with the development of towns containing fixed locations for the sale of goods and open at all times of the year, these commodity fairs began to decline. Those that survived became changed in the niche they occupied in society to perform primarily a wholesale function. They became known as 'sample fairs' and provided an opportunity for businessmen to inspect samples and design of products and to order bulk delivery of the merchandise for delivery to fixed locations in the towns and trading centres.

Two major fairs which survived these changes and still remain with us today are the Leipzig fair and the fair at Nijni-Novgorod (renamed Gorki in 1932) [5]. The Leipzig Fair has been held yearly since 1497, when it received an Imperial Sanction, although it had existed in one form or another since 1165 [6]. It flourished as an international fair not only because of its location but also because of the far-sighted command of Margrave Dietruh von Landsberg who assured the citizens of Leipzig that,

> *'All merchants, wherever they might come from, if they wish to buy or sell goods in the town, would enjoy full protection and assistance even if we are at war with the sovereigns of these merchants'.*

The Fair began as a centre of exchange in furs and skins, and grew into a general fair in consumer goods, particularly textiles from Britain (Figure 1.1). In the recent postwar period it turned to industrial equipment and is now one of the principal engineering fairs in Europe (see Chap. 14). The Gorki Fair is even older, and for many centuries business was done by barter. From 1917 to 1927 little trade was carried out but it revived in the 1930s to become the large industrial fair for Eastern Europe that it is today.

Other sample fairs took place elsewhere in Europe, in London, Birmingham, Glasgow, Lyon, Paris, Bordeaux, Milan and Turin. They succeeded mainly because of their situation on the great highways of communication and, in some cases, along the routes of holy pilgrimages.

It is these 'sample fairs' that we can recognise as a model for the international trade fairs which have become such a feature of post-industrial life in the modern world. The 'samples' became the 'exhibits', chosen as representative of the range of available products that may be ordered for business or trade.

The modern international exhibition, exposition or world fair we consider here is, however, something a little different. Here the exhibits are chosen for their novelty or uniqueness and particularly to demonstrate the achievements in invention and design of the country of origin and not simply as samples for sale (although some element of trading is always present). They were never intended as a place and time for business transactions but to provide an occasion for displaying one or more nation's products, techniques and arts in the most advantageous way to interest the general public. Accompanying this display is a measure of education and entertainment which has always been present in the English fairs, and which is an important feature of international exhibitions. The manner in which these

Figure 1.1   Selling British Textiles at the Leipzig Fair in the Middle Ages
(Reproduced by permission of E.A. Seemann, Leipzig)

additional features have entered into these occasions and the part played by technology in supporting them are considered in the next and subsequent chapters.

1.5 References

1 'Encyclopaedia Americana, vol. 10' (Grolier Inc., Connecticut, 1986, pp. 839–844)
2 'Encyclopaedia Britannica, vol. 4, (Encyclopaedia Britannica Inc., Chicago, 1995, p. 656)

3 WALFORD, C.: 'Fairs, past and present' (Cornelius Watford, London, 1883), pp. 1–11
4 DEFOE, D.: 'A tour through Great Britain' (Everyman Series, 1962, vol. 1)
5 WOODHAM, H.S.: 'Fairs' in 'Britains shop window—the BIF' (Newservice, London, 1948), pp. 12–15
6 KUCZYNSKI, K., and UNGER, M.: '800 years of the Leipzig Fair' (Leipziger Verlag, Leipzig, 1965)

Chapter 2
Early technology exhibitions

2.1 Exhibitions in Europe

Early successors to the sample fairs began to appear towards the end of the 18th century with trading fairs in Geneva, Hamburg and Prague. Vienna held national exhibitions in 1820, 1835, 1839 and 1845, the last having almost 2000 exhibitors. In Berlin and Saxony five exhibitions were held between 1822 and 1845, two displaying contributions from 3060 and 6013 exhibitors from various parts of Europe. The Berlin 1844 exhibition was a trade exhibition confined to the German states only. Other early national exhibitions are shown in Table 2.1, which includes exhibitions from Russia and the United States in the first quarter of the 19th century. In Europe, however, it was France and Britain that were to assume leadership in this activity, going about the task in quite different ways.

The French in the aftermath of the Napoleonic wars were concerned to find overseas markets for their manufactured goods, principally ceramics, tapestry and carpets (the national warehouses of Gobelin, Sevres and Savonnerie were overstocked due to the successful English blockade of the French ports over a considerable period). As a consequence the French government was sympathetic to the ideas of Marquis d'Avèze,[1] who had submitted plans for a series of expositions to help manufacturers with their marketing problems, the events to take place initially in Paris at prime locations such as the Champ de Mars and the Louvre. Evidently the British blockade rankled since Patrick Geddes notes that the organisers of the 1798 exposition '... *offered a gold medal to the exhibitor who should deal the most*

---

[1]At that time the government-appointed commissionaire for these three ex-royal manufacturers

fatal blow to English commerce' [1]! Eleven of these large national exhibitions were arranged between 1797 and 1849 under the direction of Napoleon, each one bigger than the last. The largest of these in 1849 was held at the Champs Elysées and provided plenty of space for the exhibitors to lay out their wares (Figure 2.1).

The French were the first to institute the idea of appointing an independent jury to consider the merits of the exhibits on a comparative basis. To mark their decisions a series of medals were awarded. The judging of the exhibits and award of medals was, perhaps, the most important feature of this series of French expositions. The juries were composed of men of the highest standing and their comments, as well as the medals awarded, were considered very carefully by industrialists. As a report on the 1801 exposition noted,

> *'... there is not an artist or an inventor who, once obtaining public recognition of his ability, has not found his reputation or his business largely increased'.*

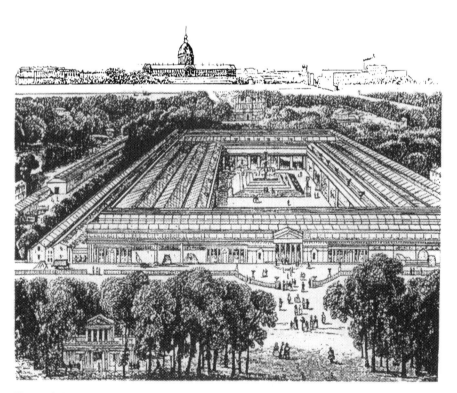

Figure 2.1   *The French Exposition Universelle 1849*
(Official Guide 'L'Exposition de Paris 1900')

This recognition was augmented at the 1801 exposition by inviting the holders of the gold medal award to dine with Napoleon, who was at that time First Consul. Table 2.2 gives some details of these exhibitions and also of the provincial exhibitions, arranged on similar lines, between 1827 and 1845. The French expositions were organised on a fairly generous scale. The 10th exposition in the Champs Elysées, for example, covered a space of 16 740 m^2 with products displayed in 40 galleries and a total avenue extent of 8–10 km. Although they often contained a number of art exhibits which would be of interest to a certain section of the population, all of these expositions were effectively what we would now call 'trade shows', displaying the products of industry with no reference to their methods of production. As a consequence the interest of exhibitors to display the possibilities of a new (and largely experimental) technology, namely the manufacture and demonstration of electrical artifacts, was minimal until the advent of the much larger French International Expositions which followed the example of the Great Exhibition in Britain in 1851.

The first industrial exhibition in Britain was held in Edinburgh in 1755 and was followed by ten more at yearly intervals and did much to stimulate Scottish industry. In England the conception of a public exhibition of industrial products was quite different from the commercial arrangements of the French and the regional view held in Scotland. The exhibition was seen as a means of educating and interesting the working classes through attendance in their hours of leisure. Indeed a certain amount of opposition existed amongst important innovators of the time—Bolton, Wedgwood, Struth and Arkwright—against industrial exhibits, on the grounds that, '... the British had eclipsed other nations in variety and excellence of their manufactures and had no need of such (foreign) ideas' [2].

2.2 Early British exhibitions

In 1828 the first attempts were made to establish an annual exhibition of a 'Catalogue of the Specimens of new and improved productions of the Artisans and Manufacturers of the United Kingdom' [3]. This initiative was due to the 'Society for the Encouragement of Arts, Manufactures and Commerce'[2]. At a meeting of the Board of Management for the Society in 1827, under the chairmanship of George Birkbeck, it was resolved to hold in Central London '... an

---

[2]'Arts' was used here in its earlier sense of, 'skill as the result of knowledge and practice' (OED)

annual exhibition of ... new and improved productions of Artisans and Manufacturers'. A Committee of Inspection was appointed which included several civil engineers, professors of the Royal Institution, and a philosophical instrument maker (Mr Watkins).

Francis Watkins was one of the early 'demonstrators of electrical effects' whose carefully constructed models and apparatus were much sought after by lecturers teaching the subjects [4]. He was one of the first to demonstrate an electric motor at the Royal Society in 1835. This consisted of a group of stationary coils facing a bar magnet, mounted on a shaft, which also carried contact breakers that sent a succession of current impulses from a battery into the coils. These impulses were so timed to give a rapid rotation to the bar magnet.

He used his motor to drive models of hammers, pumps, dredging machines, etc. and later gave an address to the Royal Society describing this and other versions of early electric motors [5].

Following the meeting of the Board, permission was obtained to hold the exhibition, which was to be known as 'The Gallery of the National Repository', in the Royal Mews, Charing Cross (see Figure 2.2)—now the site of the National Gallery—and the exhibition(s) were to be advertised as *'under the patronage of King George IV'*.

The first exhibition contained a number of Mr Watkins' models of mechanical devices and instruments for mechanical measurements. The models were of the technological innovations of the time, e.g. miniature windmills, spinning wheels, agricultural machines, model ships, etc. High technology was represented by a public showing of Lamb's Circular Proportional Slide Rule and a domestic telegraph which, disappointingly, consisted of one dial plate fixed in the parlour and the other in the kitchen, with communication between them by means of wires or chains. Later in the life of the exhibition Edward Clarke, the Dublin-trained instrument maker, was to demonstrate some of his electrical instruments at the gallery.

Public opinion of the 1828 exhibition was not good. It was called a 'toyshop' with visitors 'probing the exhibitions down and leaping upon the models', and with the Committee referred to as 'fools and paltry exhibitors' [6]. The National Repository itself was short-lived but continued for several years as 'The Museum of National Manufacturers and the Mechanical Arts', moving to a room in Leicester Square in 1833.

The difficulty facing the organisers of the Gallery has been expressed most succinctly by Altick in his excellent book, *'The shows of London'* [7], in which he remarks,

'...the staunch Victorian confidence that the public was hungry for scientific knowledge proved irreconcilable with the public's stubborn insistence that, first of all, it be amused'.

The lack of entertainment facilities in public exhibitions was noted by all concerned and the mistakes made in the National Gallery were not repeated. Indeed, as we shall see later, the careful mixing of instruction and entertainment became a feature of all subsequent public exhibitions and any failures (including financial ones) were invariably due to the lack of attention given to the choice of the correct 'mix' required for a given exhibition, taking into account local requirements as well as national aspirations.

2.3 The Adelaide Gallery

Much more successful than the Repository was the 'National Gallery of Practical Sciences' [8] set up in the Adelaide Gallery, just off the Strand, which opened at about the same time as the demise of the National Repository. The Gallery was a long room with a projecting upper level. The ground floor was dominated by a 21 m tank for the display of model ships with machines and devices alongside it (Figure 2.3). The gallery was founded in 1832 by a number of experimentalists and engineers, among them Thomas Telford the civil engineer, and supported by a wealthy philanthropist, Ralph Watson. Its objects were

> *'...to promote the adoption of whatever may be found to be comparatively superior, or relatively perfect in the arts, science, or manufacturers and .... to display specimens and models of inventions and other works ... and for the exhibition of any new application of known principles to mechanical contrivances of general utility'* [7].

It was more than an exhibition venue, however; it provided facilities for lecturers and even established a staff of 'Professors of the Gallery', of which William Sturgeon was one, to give public lectures on scientific matters. Sturgeon also made use of the Gallery as a venue for the first meetings of his London Electrical Society, the proceedings of which he edited as *Annals of Electricity*.

The Gallery attracted the interest of the American engineer, Joseph Saxton, who had arrived in London from Philadelphia in 1832. Soon after his arrival he met Jacob Perkins, an American engineer/inventor

Figure 2.2   The Gallery of the National Repository in the Royal Mews, Charing Cross (now the site of the National Gallery) (HMSO, 1862)

resident in London, who introduced him to Coleman Sellers, Thomas Gill (editor of the journal, *Technical Repository*) and to the people who were busy setting up the exhibition [9]. Saxton was commissioned to make a large permanent magnet for display in the Gallery. He was well qualified for this task, having carried out similar work at the Franklin Institute where he had published papers and gained a silver medal for his work on an Independence Clock for the city of Philadelphia.

His magnet eventually became capable of supporting a total load of 238 kg—a remarkable weight for the time. Of even more interest, however, was the use to which the magnet was put. Faraday's ideas of magnetic induction had recently been demonstrated and there was much interest at the time in finding practical ways of generating electricity from magnetism. Saxton was one of the first to devise a practical magnetoelectric machine in Britain which he exhibited at the 1833 Cambridge meeting of the British Association for the Advancement of Science, before presenting this to the Adelaide Gallery [10]. In this machine the armature coils rotated in the field of

Figure 2.3   *The National Gallery of Practical Science (The Adelaide Gallery just off the Strand, London)*
(Etching by Thomas Koerman, 1833)

a fixed magnet (Figure 2.4). He connected one armature winding to a disc and the other to a spike, which made contact with a divided mercury trough, the spike only dipping into its trough with each revolution. In this way the spike completed the circuit only when a positive voltage was present across the coils, thus producing a form of rectification of the alternating current. This method of obtaining a continuous current was common with a number of the experimental electromagnetic generators then available for purchase. He demonstrated his machine at the Gallery, incorporating the large magnet he had displayed earlier. It was a great success, producing a large spark between a pair of round terminals. This was a major accomplishment for this type of machine, although Saxton also demonstrated how the continuous direct current could be made to effect electrolysis of water. (The rate of gas production was at that time a common method of measuring the electrical power produced.) The general public may not have appreciated the niceties of this latter demonstration but they did enjoy the spark and the use to which it was put.

At intervals throughout the exhibition this was made to ignite gunpowder which, together with the noise generated by a Perkin's steam gun [11] nearby made for a noisy visit! It was a hazardous one too, as described by Armytage, who notes '... *artful snares laid for giving galvanic shocks to the unwary'!* [12]. Apparently high office was no protection, as noted by Sir Benjamin Heywood, who relates,

> '... *he saw the Duke of Wellington there, surrounded by a number of ladies: and his Grace having put his hands into the trough (containing water) used for electro-magnetic purposes, he was fastened there as completely as if his hands had been locked in a vice, and the hero of a hundred fights, the conqueror of Europe, was as helpless as an infant, under the control of that mighty agency'* [13].

The Saxton machine was demonstrated early in the life of the exhibition before an invited audience which included Michael Faraday, who had established a very close association with the new gallery. It may have been this association that brought the work of Hippolyte Pixii to the attention of the organisers.

Pixii was the first to describe a practical electro-magnetic machine to produce a unidirectional current, and he had earlier shown his machine to the public in Paris in 1832. His machine differed from Saxton's in that the field magnet revolved with respect to the coils and the machine contained a true commutator and not simply a rectifying mechanism (Figure 2.5). His commutator had originally been

Figure 2.4   Saxton's magnetoelectric machine
(Science Museum Photograph)

designed by Ampère and named by him the 'seesaw' commutator on account of its mode of operation. The rotating magnet was made to operate a cam which pushed a pair of contacts in one direction during half the revolution and the opposite direction during the other half. In this way the connections to the fixed coils could be reversed in time with the generation of the alternating voltage and so produce a continuous unidirectional current output. At this early date it was considered to be the first practical generator constructed on Faraday's principles and Faraday may well have been consulted by Pixii during its construction. Generators such as these were much in demand by experimenters of the period and the Pixii establishment in Paris was to produce four versions for sale to the Ecole Polytechnique and elsewhere. These were classified by the strength of the magnet used in their construction, the largest size containing a magnet capable of lifting a weight of 100 kg, two others lifting weights of 60 kg and 30 kg, respectively, and the smallest size designed simply to illustrate the principles involved.

By all accounts Pixii expressed in public a certain amount of pique at Saxton's claims for his machine. To settle the matter the organisers very promptly obtained a Pixii machine through the Count de Predivalli in Paris and arranged a joint demonstration of the two machines at a soireé held in the Gallery in November 1833. This was attended by Faraday, Richard Phillips (Editor of *Annals of Philosophy*), Dionysius Lardner (Professor at London University), John F. Daniell (of Daniell cell fame), William H.Pepys, Edward Turner, Joseph Henry (the American experimentalist), Moseley and several others, who reached the unanimous conclusion that the Saxton machine produced the more powerful spark [14]! The technical preparations for this demonstration were made by Edward Clarke, a demonstrator at the Gallery, who was later to produce his own machine (Figure 2.6) and to demonstrate this in 1837 before Dominique Arago, the French physicist, and other scientists at the French Institute in Paris [15]. At that time Clarke's machine was generally considered to be superior to earlier machines and established for him a lead in the production of experimental machines of this kind, a not inconsiderable financial advantage [9].

Another electrical exhibit of profound interest to the general public was the display of a number of Gymnotis, a fresh-water eel-like fish from South America, capable of giving an electric shock. These were somewhat rare specimens, seldom seen in Europe, and Faraday became very anxious to carry out some experimental work on them whilst they were in the gallery. He was later assisted by Richard Owen, a surgeon, who carried out a number of dissections of the fish under the guidance of Faraday. Some of the results of this collaborative work formed the subject of one of Faraday's Friday evening discourses at the Royal Institution in 1839 [16][3].

This period proved to be the heyday of the Gallery. Financial trouble began to be felt from 1840 onwards, and in an endeavour to remain solvent the organisers replaced the technical and scientific exhibits gradually by more and more entertainment features. By the mid-1840s the gallery had become a 'grandes soirées musicals et dansantes' and in its descent from technological display to amusement hall ceased to be of interest to the engineering and scientific community in London.

---

[3]See also a paper by D.J. Davies in the *Philos. Trans.*, 1832, p. 263, and a number of references to Faraday's notes and letters on the subject in F.A.J.L. James (Ed.): 'The correspondence of Michael Faraday, Vol. 2', IEE, London, 1993

Figure 2.5   Pixii's magnetoelectric machine
(Science Museum Photograph)

2.4 The Polytechnic Institution

The earlier successes of the Adelaide Gallery, however, made their
mark and it became obvious that a public exhibition of the efforts of

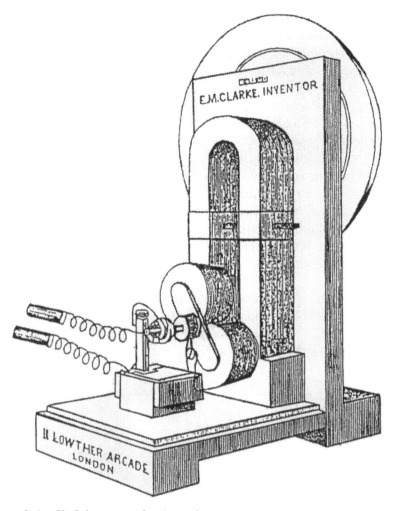

Figure 2.6   Clarke's magnetoelectric machine
(Clarke's 'Catalogue of Apparatus' 1838)

industrialists, engineers and scientists, certainly in the United Kingdom, needed to contain an element of entertainment. This was seen in nearly all the public exhibitions which followed. A number of these took place in London in the following decade. These included a 'Royal Gallery of Arts and Sciences' established in Regent Street at a venue which subsequently became the Royal Polytechnic Institution (Figure 2.7). The Polytechnic Gallery was in operation at the same time as the Adelaide Gallery, having been opened in 1838. It was, however, to outlive the Adelaide Gallery due to the sounder management of its founder, Sir George Cayley, physicist, experimental engineer and frequent contributor to the *Mechanics Magazine* [17]. It was at this gallery that Alexander Bain was to exhibit in 1842 his Printing Telegraph, an invention significantly ahead of its time [18]. In founding the Royal Gallery of Arts and Sciences, Cayley was joined by a group of members of Parliament and practical scientists, including a Professor Bachhoffner, who was involved in planning and day by day operations. The Royal Polytechnic Institution received its Royal Charter in 1839 and from its inception established a firm educational basis for training of technologists and engineers. Its exhibition hall, a gallery 36 m long and 12 m wide, exceeded in size that of the Adelaide Gallery, but its contents were a miscellaneous collection with few of the electrical and engineering attractions of this latter gallery. It did, however, excel in its range of popular lectures—it included a 500 seat lecture hall on the premises. The lecturers made full use of the 'dissolving view' slide projector apparatus then just coming into use. The subjects were chosen from popular themes of the day, including talks on the electric telegraph, atmospheric electricity and mining machines. John Henry Pepper was its chief lecturer from 1848 onwards and was inventive in combining entertainment with scientific demonstrations. In a demonstration of the principles of acoustics he

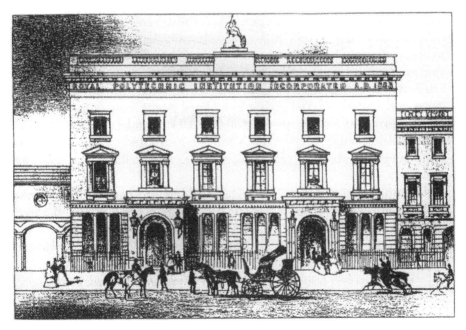

Figure 2.7   Royal Polytechnic Institution
(Lithograph by G.J. Cox, 1838)

referred to 'the beautiful experiments of Professor Wheatstone' by which four harps play 'without visible hands as the sounds are conducted to them by rods from instruments played by performers who are placed several floors beneath the lecture room' [19].

2.5 The Royal Panopticon of Science and Art

In the 1850s a permanent exhibition was set up in Leicester Square by E. Marmaduke Clarke, its resident Managing Director, the same Edward Clarke who followed Saxton and Pixii in the successful production of electro-magnetic machines. This was housed in a large building, built especially for the purpose. It had an imposing 32 m frontage and a rotunda 28 m high, its interior being highly decorated in the Victorian manner with embellishments including a grand organ and a complex steam heating system for the public rooms [20][4].

Two daily exhibitions were held, one in the morning for scientific demonstrations and in the evening a more popular form of education, blending artistic entertainment with instruction. A permanent exhibit in the rotunda was contained in a large cistern, big enough to hold a diver in full rig, observed through large glass panels. The diver carried a subaqueous lantern comprising a hydro-oxygen lime light and with this illumination demonstrated the process of raising sunken artifices from the bed of the tank through the use of an inflatable balloon. Another demonstration was attributable to Michael Faraday and concerned the liquifaction of gases, particularly that of carbonic-acid gas. Apart from demonstration equipment the exhibition contained a number of engineering exhibits including several machines contributed by Whitworth, his bolt screw production machine, a lathe, and drilling and planing machines.

A number of Clarke's electrical and electromagnetic machines were shown. One of these, a static electricity generator, was claimed to be

> *'the largest ever constructed'* and the organisers '... *were influenced by a conviction of the danger with which a display of the powers of the hydro-electric machine is ever fraught, not only to the*

---

[4]'Panopticon'—Bentham's name for a proposed 'form of prison of circular shape having cells built round a central well, whence the warders could at all times see the prisoners (OED)'. If we read 'warders' as 'visitors' and 'prisoners' as 'exhibits', then this describes Clarke's building very well

operator, but to the entire audience ... With the gigantic plate machine, the experiments will be on the brilliant scale, but totally divested of danger'.

A sketch made at the time is shown in Figure 2.8. In view of the claim that the rotating glass plate is over 3 m in diameter, the scale does look correct. It was worked by steam power from the basement but the writer of the Handbook does not give any further technical details [21][5]. A demonstration of Duboscq's arc light situated high in the rotunda was frequently given. This was energised by a 'galvanic battery' and the separation of the carbons arranged by clockwork. The telegraph equipment of J.W. Brett was shown, following his recent successful installation of the crosschannel cable connecting Britain with France in which he used the new Gutta Percha insulation (Figure 2.9).

2.6 Provincial exhibitions in Britain

Other semiprivate exhibitions at that time were the 1845 Manchester School of Design's 'Exposition of Arts and Manufactures', containing

> *'...products of the British loom, British machinery, and of the first and most perfect of all machines, the human hand'.*

In the same year in London a Free Trade Bazaar in Covent Garden had similar aims to present a view of manufacturing England.

Exhibitions of machinery models were given periodically at the rooms of the Royal Society of Arts in John Adam Street, London (of which more later); and in Leicester an exhibition was held in 1840 at which Uriah Clark of Leicester demonstrated his 'electromagnetic carriage' on a circular railway set up inside the hall. His machine worked on a pivoted lever subject to a series of short sharp pulls by the action of an electromagnet [22]. A number of these nonrotational machines were designed at about this time and several shown at the 1851 exhibition, considered in Chapter 5. National exhibitions were also arranged at Bingley Hall, Birmingham, in 1839 and 1849, held in connection with meetings of the British Association for the Advancement of Science. The 1839 meeting saw the first demonstration of George Knight's electromotive engine, later to be shown at the 1851 exhibition.

---

[5]A similar large Wimshurst machine featured later at the Vienna exposition of 1883. This contained a glass disc of 2 m in diameter and was also driven by a steam engine [21]

Figure 2.8   Clarke's static electricity generator
(White, 'Handbook of the Royal Panopticon', 1854)

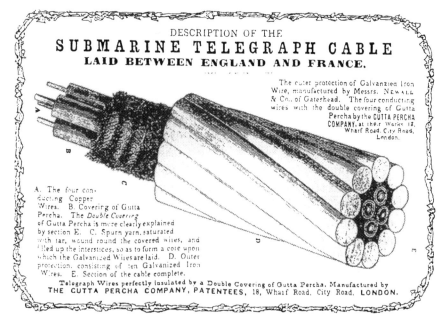

Figure 2.9   A submarine telegraph cable
(White, 'Handbook of the Royal Panopticon', 1854)

Table 2.1   Major technological exhibitions, 1750–2005

(Excluded are Trade Fairs and Exhibitions which do not contain any technology exhibits (e.g. art or agricultural exhibitions). Included are proposed world fairs up to the year 2005)

Year	Title	Days open	Location	No. of visitors (thousands)
1756	Earliest RSA Exhibition		UK	
1789	Geneva		Switzerland	
1790	Hamburg		Germany	
1791	Prague National Exhibition		Czechoslovakia	
1797	Paris National Exhibition		France	
1798	1st Paris Exposition		France	
1801	2nd Paris Exposition		France	
1802	3rd Paris Exposition		France	
1806	4th Paris Exposition		France	
1819	5th Paris Exposition		France	
1823	6th Paris Exposition		France	
1827	7th Paris Exposition		France	
1828	National Repository Exchange, London	about 3 years	UK	
1833	Adelaide Gallery, London	semipermanent	UK	>500
1834	8th Paris Exposition		France	
1838	Polytechnic Gallery Exhibition, London	semipermanent	UK	
1839	9th Paris Exposition		France	
1844	RSA 'Grand Exhibition', London	1	UK	2
1844	10th Paris Exposition		France	
1845	Free Trade Bazaar, London	7	UK	73
1849	RSA Manufacturers Exhibition, Covent Garden	7	UK	100
1849	11th Paris Exposition		France	
1850	Dublin National Exhibition	182	Ireland	300
1851	1st Great Exhibition, London	141	UK	6039
1852	Cork Exhibition	182	Ireland	589
1853	Dublin World Fair, 1st International Exhibition	171	Ireland	1160
1853	New York 'Crystal Palace' World Fair	139	USA	1250
1855	Paris International Exposition	200	France	5162
1857	Manchester		UK	1337
1858	Philadelphia		USA	100
1862	Paris International Exposition		France	
1862	2nd Great Exhibition, London	171	UK	6211
1865	Dublin 2nd International Exhibition	151	Ireland	932
1867	Paris International Exposition	217	France	11000
1871	London 1st 'Selected Works' Exhibition	150	UK	1142
1872	London 2nd 'Selected Works' Exhibition	165	UK	647
1873	London 3rd 'Selected Works' Exhibition	210	UK	500

Table 2.1   continued

Year	Title	Days open	Location	No. of visitors (thousands)
1873	Vienna International Exhibition	186	Austria	7254
1874	London 4th 'Selected Works' Exhibition	214	UK	467
1874	Marseille Modern Inventions		France	
1874	Philadelphia Franklin Society Exhibition	182	USA	268
1876	London Loan Collection of Scientific Apparatus	182	UK	213
1876	Paris Applications of Electricity		France	
1876	Philadelphia Centennial Exhibition	160	USA	9857
1878	Paris International Exposition	190	France	16000
1879–80	Sidney International Exhibition	214	Australia	1117
1880	London Lighting Exhibition, Alexandra Palace		UK	
1880–81	Melbourne International Exhibition	210	Australia	1459
1880	Glasgow Electrical Exhibition	33	UK	40
1881	1st London Electrical Exhibition	40	UK	
1881	Atlanta	100	USA	1113
1881	Paris Electrical Exhibition	182	France	
1882	Munich Electrical Exhibition	60	Germany	
1882	Dublin		Ireland	261
1883	Westminster Aquarium Exhibition, London		UK	
1883	Manchester Electrical Exhibition		UK	
1883	Amsterdam	153	Holland	1000
1883	Vienna Electrical Exhibition	108	Austria	
1883	London Fisheries Exhibition	182	UK	500
1883–4	Boston	120	USA	300
1883–4	Calcutta	96	India	1000
1883–7	Louisville Southern Exposition	61	USA	1000
1884	London Health and Education Exhibition	182	UK	
1884	Philadelphia Electrical Exhibition	40	USA	285
1884–5	New Orleans Industrial Fair	165	USA	3500
1885	London Inventions Exhibition	182	UK	3485
1885	Antwerp	180	Belgium	3500
1886	Edinburgh Jubilee Exhibition	177	UK	2729
1886	Newcastle Jubilee Exhibition	149	UK	2092
1886	London Colonial and Indian Exhibition	188	UK	5551
1887	Manchester Jubilee Exhibition	270	UK	4509
1887	Liverpool Jubilee Exhibition	182	UK	4156
1887	Saltaire Jubilee Exhibition	153	UK	823
1887	London American Exhibition	182	UK	
1887–8	Adelaide Jubilee Exhibition	210	Australia	790
1888	London Italian Exhibition	182	UK	

Table 2.1   continued

Year	Title	Days open	Location	No. of visitors (thousands)
1888	Barcelona International Exposition	182	Spain	1227
1888	Irish International Exhibition (London)		UK	
1888	Brussels Science and Industrial Exposition	182	Belgium	
1888	Glasgow International Exhibition	172	UK	6039
1888–9	Melbourne Centennial Exhibition	184	Australia	2200
1889	Paris Centennial Exhibition	165	France	32350
1889–90	Dunedin	125	New Zealand	618
1890	London French Exhibition	143	UK	1300
1891	London German Exhibition	150	UK	6700
1891	Kingston	80	Jamaica	304
1891	Frankfurt Electrotechnical Exhibition	150	Germany	1200
1892	2nd London Electrical Exhibition		UK	
1893	Chicago Columbian World Fair	188	USA	27000
1894	Antwerp	182	Belgium	3000
1894–5	Californian Midwinter Exposition, San Francisco	155	USA	1356
1894–5	Tasmanian International Exhibition, Hobart	182	Australia	290
1895	Atlanta	134	USA	6000
1897	London American Exhibition	151	UK	2200
1897	Brussels	193	Belgium	6000
1898	Omaha Trans–Missouri Exposition	180	USA	2614
1900	Paris International Exposition	213	France	48000
1901	Buffalo Pan-American Exhibition	190	USA	8120
1901	Glasgow International Exhibition	191	UK	11560
1902–3	Turin	227	Italy	
1902–3	Hanoi	100	Viet-nahm	4000
1904	St Louis Louisana Purchase Exhibition	198	USA	19695
1905	Liège	182	Belgium	6143
1905	Portland	165	USA	2554
1906–7	Milan	182	Italy	5500
1906–7	Christchurch	182	New Zealand	2000
1907	Dublin International Exhibition	182	Ireland	2751
1907	Jamestown Tercentennial Exhibition	182	USA	2551
1908	London Franco–British Exhibition	150	UK	8400
1909	Seattle	182	USA	3741
1910	Brussels	212	Belgium	13000
1911	London Festival of Empire	167	UK	
1911	Glasgow International Exhibition	182	UK	9500
1911	Turin Industrial and Labour Exhibition	135	Italy	4013
1913	Ghent	230	Belgium	11000
1915	San Francisco Pan-Pacific Exhibition	280	USA	18876

Table 2.1   continued

Year	Title	Days open	Location	No. of visitors (thousands)
1915–6	San Diego	360	USA	3800
1918	New York Bronx International Exposition		USA	
1922–3	Rio de Janeiro	365	Brazil	3626
1923	Calcutta International Exhibition	106	India	1000
1924–5	London British Empire Exhibition, Wembley	360	UK	27000
2924–5	Dunedin New Zealand & South Seas Exhibition	165	New Zealand	3200
1925	Paris International Exposition	180	France	5853
1926	Philadelphia Sesqui-Centennial Exposition	180	USA	6400
1928	Long Beach Pacific South-west Exposition	36	USA	1100
1929	Barcelona International Exposition	180	Spain	15000
1930	Seville Ibero–American Exposition	72	Spain	1500
1930	Antwerp/Liège	180	Belgium	>500
1931	Paris Colonial Exposition	180	France	33500
1933–4	Chicago 'Century of Progress' Exhibition	360	USA	48800
1935	Brussels International Exposition	180	Belgium	20000
1935–6	San Diego Californian–Pacific Exposition	380	USA	6790
1936	Texas Centennial Central Exposition, Dallas	150	USA	
1936–7	Johannesburg Empire Exhibition	120	South Africa	1500
1937	Düsseldorf 'Nation at Work' Exhibition	160	Germany	5000
1937	Paris—Arts and Techniques Exposition	180	France	34000
1938	Glasgow British Empire Exhibition	180	UK	13500
1938	Marseille International Exposition		France	
1939–40	San Francisco Golden Gate Exposition	365	USA	17041
1939–40	New York World Fair	335	USA	45000
1939–40	Wellington Centennial Exhibition		New Zealand	2641
1946	London 'Britain can make it' Exhibition	160	UK	1500
1951	RSA 'Exhibition of Exhibitions'	160	UK	
1951	London Festival of Britain	150	UK	8500
1958	Brussels World Fair	182	Belgium	41500
1961	Italia '61, Exposition, Turin	182	Italy	
1962	Seattle Century 21 Exposition	182	USA	9700
1964–5	New York World Fair	360	USA	51600
1967	Montreal Expo'67	182	Canada	50300
1967	Fairbanks Alaska '67 Centennial Exposition		USA	
1968	San Antonio Hemisfair	180	USA	6384

Table 2.1   continued

Year	Title	Days open	Location	No. of visitors (thousands)
1970	Osaka Expo '70	182	Japan	64200
1974	Spokane, Washington Expo '74	182	USA	5600
1975–6	Okinawa International Ocean Exposition	182	Japan	3480
1982	Knoxville International Energy Exposition	182	USA	11000
1984	Louisiana New Orleans World Exposition	182	USA	7300
1985	Tsukuba Expo'85	182	Japan	20000
1986	Vancouver Expo'86	182	Canada	22000
1988	Glasgow	150	UK	4346
1988	Brisbane Expo'88	180	Australia	15700
1992	Seville Expo'92	182	Spain	1896
1993	Taejon Expo'93	182	Korea	10000
1995	Internet World's Fair	N/A	Cyberspace	N/A
1996	Internet World Exposition	N/A	Cyberspace	N/A
1997*	Hong Kong Expo'97		China	
1997*	Stockholm Exposition	120	Sweden	2500*
1998*	Lisbon Expo'98	120	Spain	9000*
1999*	San Francisco World Fair	182	USA	25000*
2000*	Beijing Exposition		China	
2000*	Denver World Exposition	182	USA	16000*
2000*	Hannover Expo'2000	160	Germany	40000*
2000*	Millenium Exhibition	182	UK	15000*
2001*	Swiss National Exposition	182	Switzerland	12000*
2005*	Aichi World Exposition	182	Japan	25000*
2005*	Calgary Expo'2005	182	Canada	12000*

* = proposed/estimated
Other smaller and provincial exhibitions having some technical content were held in:

Munich (1818), Vienna (1820), Ghent (1820), Berlin (1822), Stockholm (1823), Saxony (1824), Dresden (1824), Tournai (1824), Haaelwm (1825), Dublin (1826), Berlin (1827), Madrid (1827), New York (1828), Moscow (1829), St.Petersburg (1829), Dublin (1829), Dublin (1833), Berlin (1834), Lausanne (1834), Hannover (1835), Vienna (1835), Brussels (1835), Lausanne (1837), Dublin (1837), Vienna (1839), Brussels (1840), Washington (1844), Stockholm (1844), Berlin (1844), Lisbon (1844), Dublin (1845), Saxony (1845), Vienna (1845), Zurich (1848), Moscow (1849), Birmingham (1849), Madrid (1850), Munich (1854), London Education (1854), Madrid (1854), Christiaana (1854), Bern (1857), Lausanne (1857), New York (1858), Hannover (1859), Melbourne (1861), Haarlem (1861), Melbourne Victorian (1861), Florence (1862), Rome (1862), Constantinople (1863), Amsterdam (1864), 1st London Working Classes (1864), Dunedin (1865), Vienna (1865), Stettin (1865), Philadelphia (1865), 2nd London Working Classes (1866), Stockholm (1866), London Aeronautics (1868), 3rd London Working Classes (1869), St Petersburg (1870), Melbourne (1872), Moscow Polytechnic (1872), Copenhagen (1872), Dublin (1872), Chicago (1873), Cincinnati (1874), Melbourne (1875), Ghent (1877), Cape of Good Hope (1877), Berlin (1879), Milan (1879), Turin (1880), Brussels (1880), Adelaide (1881), Christchurch (1882), Moscow (1882), Bordeaux (1882), Ehrenfeld (1882), Frankfurt (1882), Paris Railway Appliances (1883), Chicago Railway (1883), Caraccus (1883), Zurich (1883), Prague Electrical (1883), Turin (1884), Nice (1884), Toronto (1884), Wellington (1885), Budapest (1885), Königsberg (1885), Parma (1887), Lille (1889), Edinburgh Electrical (1890), Madrid (1892), Moscow (1892), Berlin (1892), Kimberley (1893), Madrid (1894–1896), Rome (1895), Kyoto (1895), Nashville (1897), Stockholm (1897), Brisbane (1897), Guatemala (1897), London Universal (1898), Como Electrical (1899), Charleston (1901), Düsseldorf (1902), Tokyo (1907), Wellington (1911), Port-au-Prince (1949), Lausanne (1963).

Table 2.2   French exhibitions before 1851

Year	Title	Location	Exhibits
1797	National Exhibition	Paris	
1798	1st French Exposition	Champ de Mars	110
1801	2nd French Exposition	Paris	229
1802	3rd French Exposition	Paris	540
1806	4th French Exposition	Paris	1422
1819	5th French Exposition	Paris	1662
1823	6th French Exposition	Louvre	1648
1824	Tournai		
1827	Nantes		
1827	7th French Exposition	Louvre	1795
1834	8th French Exposition	Paris	2447
1835	Lille		
1835	Bordeaux		
1836	Toulouse		
1836	Dijon		
1839	9th French Exposition	Champs Elysées	3281
1844	10th French Exposition	Champs Elysées	3960
1845	Bordeaux		
1849	11th French Exposition	Champs Elysées	4494

2.7 References

1 GEDDES, P.: "Industrial exhibitions and modern progress' (Douglas Printers, Edinburgh, 1887)
2 HOLLINSHEAD, J.: 'A concise history of the International Exhibition of 1862' (HMSO, London, 1862)
3 'The Gallery of the National Repository'. Report, RSA Library No. 83638, 1828
4 GEE, B.: 'Electromagnetic engines', in HOLLISTER-SHORT, G., and JAMES, F. (Eds.): 'History of Technology, Vol. 13' (Mansell, London, 1991), pp. 41–72
5 'Watkin's motor', *Philos. Trans. R. Soc.*, 1835, **7**, p. 10
6 'The International Exchange of 1862—illustrated catalogue of the Industrial Department, British Division, Vol. 1'. Report, RSA Library, 1862
7 ALTICK, R.D.: 'The Shows of London' (Belknap Press of Harvard University, Cambridge, MA, USA, 1978)
8 'National Gallery of Practical Sciences, Adelaide Street', *Lit. Gaz.*, 16th Nov. 1833, p. 730
9 GEE, B.: 'The early development of the magneto-electric machine', *Ann. Sci.*, 1993, **50**, pp. 101–133
10 *Ann. Electr.*, 1837, pp. 145–155
11 BATHE, G.: 'Jacob Perkins: his inventions, his times and contemporaries' (Historical Soc. of Pennsylvania, Philadelphia, 1943)
12 ARMYTAGE, W.H.G.: 'A social history of engineering' (Faber and Faber, Cambridge MA, 1970), p. 146
13 'Report of the Directors of the Manchester Mechanics Institute' (W. Simpson, Manchester, 1838), p. 56
14 *Philos. Mag.*, (1835), **6**, p. 239; *Lit. Gaz.*, 1833 (878), p. 730
15 STURGEON, W.: *Ann. Electr.*, 1837, pp. 145–155

16 FARADAY, M.: 'Faraday's Notes'. Royal Institution manuscript F4, 1839 (110), p. 19
17 PRITCHARD, L.J.: 'Sir George Cayley: The inventor of the aeroplane' (Max Parrish, London, 1961), pp. 125–127
18 BURNS, R.W.: 'Alexander Bain, a most ingenious and meritorious inventor', *Eng. Sci. Educ. J.*, 1993, **2** (2), pp. 85–93
19 Illustrated London News, 3rd February 1856, pp. 117–118
20 WHITE, W.: 'Illustrated Handbook of the Royal Panopticon of Science and Art—an Institution for Scientific Exhibitions and for Promoting Discoveries in Arts and Manufactures'. (Guildhall Library, London, 1854)
21 'Vienna Exhibition of 1883', *Electr. Rev.*, 1885, **6**, 2 Feb. 1885, p. 3
22 *Ann. Electr.*, 1840, **5**, pp. 33–34, plate 1; pp. 304–305

Chapter 3
Role of the mechanics' institutes

3.1 The mechanics' institutes

The mechanics' institutes were important for the future of technology exhibitions in Britain particularly for the interest they engendered in the public, shown by the impressive attendance figures for their exhibitions in the 1840s. It is at this level that they can be seen as a stimulant to later discussions at the Society of Arts on the planning for the Great Exhibition of 1851. The organisation of the mechanics' institutes exhibitions lay in the hands of the Directors, members and committees of the individual institutes throughout Britain. Although supported by local manufacturers, popular educators and professional educationalists, such as George Birkbeck, they failed to attract directly the interest of the scientific or engineering communities, which were drawn towards the professional organisations: The Society of Arts, The Royal Institution, The Royal Society, the Universities and later to the Institutes of Mechanical and Civil Engineers. There is no doubt, however, that the mechanics' institutes stimulated a keen interest in the artisan public and their friends to attend technology exhibitions in Britain, and it was their numbers that provided a sound nucleus of support for the national and international exhibitions that followed in the latter half of the 19th century [1].

The origin and subsequent evolution of the mechanics' institutes is directly associated with the Industrial Revolution. The constant stream of new inventions and discoveries then occurring needed a new breed of workman familiar with at least the ideas of scientific progress. There was also a growing popular interest amongst the public in science and the development of institutions to provide scientific and mechanical engineering instruction at a level that could be understood by the mechanics of the day having little or no supportive academic background[1]. Finally, and most importantly, a basic desire existed among the manual workers, particularly the craftsmen and guildsmen, to *'better themselves'* through their own efforts. This last was a motive constantly repeated in much of the popular literature of the time (e.g. *'Lives of Engineers'* (1862) and *'Self-help'* (1859), both by Samuel Smiles), with constant reference to the names of successful working-class inventors such as Arkwright, Brindley, Franklin, Rennie, Smeaton and Watt. It is of interest to note that the encouraging material included in *'Self-help'* was presented by Samuel Smiles himself, on one occasion, to about a hundred working men in Leeds, who soon afterwards, entirely on their own initiative, set up an evening school with lectures on topics considered needed for *'mutual improvement'* [2]. Thus the notion of an Institute to provide at least some of this kind of assistance for the working man had been expressed by a number of people in the first few decades of the 19th century, sometimes taking the form of 'mechanics classes' or free admission to scientific lectures held at existing institutions and occasionally organised by local groups of workers themselves.

An early supporter for this work was John Anderson on his appointment to the chair of Natural Philosophy at Glasgow University in 1760. From then on to his death in 1796 he gave lectures which he invited artisans to attend, *'... in their working clothes'*. His work was continued at Glasgow, and later in London by Dr George Birkbeck, and prompted proposals there for an Institute to be run by the workers themselves. These views were expressed in the popular *Mechanics' Magazine* for a society which would be concerned with the leisure and education of a new working class, the skilled mechanics and artisans who provided the work force of the new industries [3]. Similar proposals were made in other quarters and were taken up by George Birkbeck, who had been responsible for a series of mechanics' classes in Glasgow, which went some way to satisfying the worker's needs. When asked to become a patron of a new body to be called the Glasgow Mechanics' Institute in 1823, then being organised by a group of dissenter students from the mechanics' class at Anderson's Institute, he responded with enthusiasm [4]. This was to be the first of the mechanics' institutes known by that name, although rival claims are made for Edinburgh and Birmingham who had established similar

---

[1]They were, however, literate. In 1841 one third of the men and almost half of the women marrying in England and Wales signed the marriage register with a mark

organisations a few years earlier. Birkbeck went on to co-operate with J.C. Robertson, the editor of the *Mechanics' Magazine*, who had made a proposal for a similar organisation for London in his magazine a year earlier, and they jointly founded the London Mechanics' Institute, also in 1823. The objects of the new institution were briefly stated as,

> 'The instruction of the members in the Arts they practise, and in other branches of scientific and useful Knowledge' [5].

The London Mechanics' Institute was to play a leading part in the National Repository and the Museum of National Manufacturers and the Mechanical Arts, noted earlier; indeed the Director of this latter exhibition, Charles Toplis, was at that time, Vice-President of the London Mechanics' Institute. From this date the movement spread rapidly, supported by George Birkbeck (until his death in 1841) and local industrialists, so that by the 1840s there were more than 50 000 members in about 200 mechanics' institutes in provincial towns throughout the country. By the time of the Great Exhibition in 1851 this number had increased to almost 700 (a figure which includes similar organisations, such as the various lyceums, and literary and philosophical institutes [4]). Figure 3.1 shows the distribution of the mechanics' institutes in Britain as they appeared in 1841.

3.2 Exhibitions at the mechanics' institutes

Exhibitions formed a major part of the activities of these institutes and, following visits of provincial members to London in the 1820s and 1830s to see the various gallery exhibitions, the idea of holding similar exhibitions in their own localities gained ground. Collection of suitable objects for display was no doubt helped by the loans and donations of equipment made by local industrialists. A notable engineering exhibit, donated locally for the Manchester Mechanics' Institute, was a lathe screw, 7 cm in diameter and 9 m long!

The first exhibition was organised in Lancashire by the small Louth Mechanics' Institute in 1835 [6]. Other exhibitions quickly followed, in Manchester, Leeds, Sheffield and other manufacturing towns. They were extraordinarily successful in attracting visitors (usually at 6d ($2\frac{1}{2}$p) a time). In Manchester alone four exhibitions between 1837 and 1842 attracted more than 300 000 people. The Leeds exhibition in 1840 admitted 200 000, with similar attendances at Liverpool, Sheffield, Derby and other towns. At most of these large town

Figure 3.1   The distribution of Mechanics' Institutes in Britain in 1841
(Reproduced with permission from Kelly, 'George Birkbeck—Pioneer of Adult Education', Liverpool University Press, 1957)

locations the military and police forces were admitted free (probably considered as a wise precaution at the time!) together with upwards of 20 000 children attending from the different charity schools. In all several million visitors must have visited these exhibitions—to such an extent that publicans were concerned that their customers were said to be forsaking the public house to visit instead the Mechanics' Institute exhibitions with their wives and families! [7]. This popularity amongst artisans and their families often came as a surprise to the organisers, and in Leeds, for example, the committee found it necessary to close the doors twice against newcomers and the *Leeds Mercury* in July 1839 noted that,

> *'From about the middle of each day the scene is very animated, but in the evening—between the hours of seven and ten, the rooms ... are crowded that they would be almost unbearable, but that everybody seems too much intent on the objects which solicit their notice to regard personal inconvenience arising from the heat'.*

At its peak 20 or more significant exhibitions would be given throughout the country during the course of a year. Table 3.1 lists some of the major exhibitions arranged by the mechanics' institutes over the most productive period, 1837 to 1845. The fact that this period corresponds with the 'railway boom' of the early 19th century is no coincidence (see Figure 3.2). A writer in a Derby newspaper reported that,

> *'The facilities offered to the public by the opening of the railway to Nottingham, have doubtless contributed to this great influx of visitors (to the Nottingham Mechanics' Institute Exhibition)'.*

Table 3.1   Major exhibitions arranged by the Mechanics' Institute

1837	Manchester
1838	Manchester (2nd exhibition), Liverpool, Potteries, Newcastle, Sunderland
1839	Salford, Leeds, Sheffield, Potteries (2nd exhibition), Derby, Birmingham, Preston, Macclesfield, Sowerby Bridge
1840	Manchester (3rd exhibition), Wigan, Preston (2nd exhibition), Sheffield (2nd exhibition), Halifax, Huddersfield, Nottingham, Norwich, Stockport, Stroud, Leeds (2nd exhibition), Leicester, Liverpool, Beccles, Rippon, Birmingham (2nd exhibition), Bradford, Newcastle (2nd exhibition), Oldham, Salford (2nd exhibition)
1842	Liverpool (2nd exhibition), Manchester (4th exhibition)
1843	Leeds (3rd exhibition), Derby (2nd exhibition)
1844	Liverpool (4th exhibition)

British Railway System 1845	British Railway System 1851

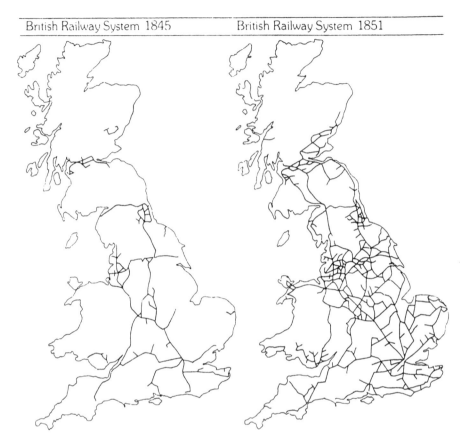

Figure 3.2   Growth of the railways in the 1840s

Mechanics' institutes in various parts of the country, newly opened to the railway, were active in associating railway excursions with a visit to their exhibitions. An excursion organised by the Nottingham Institute to visit the exhibition at neighbouring Leicester was carried out on a grand scale, and the *Leicester Journal* was to report,

> '... *upwards of 2000 persons assembled, and arrayed in their gayest garments, with banners, and handkerchifs, and hats waving in the air, they were shortly afterwards in motion. The enormous train, consisting of 65 carriages, and measuring upwards of a quarter of a mile in length, proceeded along the Trent meadows, amidst the admiration and acclamations of thousands who had come out to witness this prodigy of steam and exult in, although they could not join, this new movement which railways have given to social intercourse*' [8].

Whilst the educational value of these exhibitions and the entertainment they provided to members and their guests were much appreciated, a major factor in the enthusiastic acceptance of these events by the mechanics' institutes committees was the profit they made. This could be quite appreciable. Bradford made a profit of £700 by holding one exhibition and the five exhibitions arranged by Manchester realised over £5000, with a similar figure for Liverpool for four exhibitions. The extra finances were used by the institutes mainly to pay for buildings containing much-needed lecture rooms, classrooms and members' rooms.

The number of exhibitions declined after the 1840s—some mechanics' institutes, having cleared their debts by the exhibition profits, did not attempt to repeat their success. The Manchester Mechanics' Institute gave priority to classes and lectures (exhibitions were often staged in classrooms, blocking their use for lecturing purposes). West Riding reported in 1840 that, *'the novelty of this class of entertainment had passed away...'*. Others opined that *'... the state of trade was not in their favour'*. They were not abandoned without opposition however, and one focus of dissent, addressed as a public letter, *'... To the Working Men of Manchester...'*, calculated that if the exhibitions could be continued they would provide enough funding for, *'...a botanical garden, a natural history museum and a gallery of sciences for your city'* [9]— anticipating some ten years later the distribution of profits for the Great Exhibition in London in 1851 (see Chap. 5). The exhibitions had been valuable, however, despite their limited history; the public interest had been engendered and the stage was set for future public exhibitions on similar lines.

The exhibits themselves were a conglomeration of fine arts, natural history and mechanical models. Some of these latter were very impressive and noisy, following the example set by the Adelaide Gallery. On occasion suitable items were, in fact, borrowed from the Gallery for exhibition in a provincial institute. One quite large example was a demonstration of a series of models of the Atmospheric Railway (a Victorian engineering invention of great interest at the time [10]) by a Mr Hay of the Portsmouth Athenaeum, lecturing at the Manchester Mechanics' Institute. The models were 17 m long and, *'... a small carriage, with a youthful passenger will, in connection with the experiments, be propelled along a line of railway by the pressure of the atmosphere solely'* [11]. A full-sized version of the Atmosphere Railway was constructed in 1865 to link the upper and lower levels of the Crystal Palace railway station after the Palace was transferred to Sydenham to become a permanent exhibition site.

Other working mechanical models included the Jacquard loom exhibited at several mechanics' institutes in the north of England (shown previously for the first time at the second French exposition in 1801), a letter press and other models powered by miniature steam engines. Often these model working engines produced actual products, which could then be sold to visitors. Perhaps on this account manufacturers were often pressed to loan a machine (with operator) to a given mechanics' institute, with this request repeated (often in the same year) by other institutes, so that manufacturers had difficulty in satisfying all their requests. Often the exhibits were linked to lectures given within the institute. Drawings and artifacts illustrating statics and dynamics, hydrodynamics, pneumatics, heat, light, electricity, astronomy, geodesy and chemistry were some of the subjects discussed over the year. The scientific apparatus exhibited often included microscopes and telescopes together with a few electrical items such as powerful electromagnets and the ubiquitous 'galvanic machine', which we now recognise as a 'shocking coil', having reputed medical advantages [12], but used as a source of much merriment. Figure 3.3 shows a typical example of a 'family electromedical apparatus' then coming on to the commercial market.

3.3 Manchester Mechanics' Institute and its exhibitions

The Manchester Mechanics' Institute in Cooper Street (Figure 3.4) was typical of the larger and successful institutes in the north of England. It was situated at the heart of the Industrial Revolution, with a local population of artisans employed in a wide variety of engineering trades and whose numbers increased yearly. (The population of Manchester had increased from 84 000 in 1801 to 460 000 in 1861.) It also had the strong support of local industrialists who had good reason to encourage their technical workers to become more proficient.

The Institute was founded in 1825. The leading figures bringing this about were William Fairbairn, already a famous engineer, Thomas Hopkins, an alderman, and Richard Roberts, the inventor of the self-acting mule. They were joined by a number of business men including Benjamin Heywood, the banker. Shortly afterwards,

'... *eight gentlemen connected with the trade and manufactures of the town, liberally advanced sums of £500 each for the purchase of land and the erection of a suitable building'* [13].

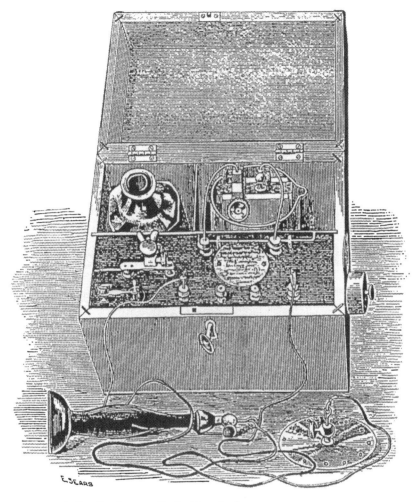

Figure 3.3   A 'family electromedical apparatus'
(Official Catalogue of the Great Exhibition, 1851)

This was in fact, the first building constructed specifically as a mechanics' institute. By 1843, Sir Benjamin Heywood, FRS, then current President, was able to report that,

> '... *the debt incurred has since been materially reduced partly by good management ...and partly due to the periodical Exhibitions within its walls, which have afforded so much instruction and delight to the working classes in Manchester and other towns*'.

How much delight is encapsulated by a certain Mr L.M. Hayes, a Manchester citizen, writing of his recollections of childhood in the 1850s:

THE MANCHESTER MECHANICS' INSTITUTION
Erected in 1825
From the Jubilee Book of the Manchester College of Technology

Figure 3.4   The Manchester Mechanics' Institute in Cooper Street
(Reproduced with permission from Kelly, 'George Birkbeck—Pioneer of
Adult Education', Liverpool University Press, 1957)

*'The old Mechanics' Institute ... in Cooper Street in the building
opposite the present Town Hall ... was a place started for the
improvement, intellectually and socially, of the working and middle-
classes; subscription 5s per quarter. Exhibitions and entertainments
of an educational and interesting character were from time to time
held in it, and as we lived near at hand it was a kind of rendezvous
for the young people of the families round about....my family had a
season ticket, enabling me to go whenever I wished, and I wished
very often, as did also my young friends. I have a strong conviction
that the officials would often have preferred our room to our
company, as we practically turned the building into a playground,
and rambled over the whole place ... and were nearly always
somewhere where we had no right to be, and in one kind of mischief
or another.... There was a Centrifugal, or 'Loop the Loop' Railway,*

... much patronised by those who had a weakness for being sent spinning round and round like a human pinwheel.... Another object of interest was the Galvanic Battery, which gave you all kinds of shocks, strong or mild as desired. One time when boasting to my companion that I could take the strongest indicator point shock, I illustrated my ability to do so by taking hold of the handles with both hands after pushing the needle to full strength. When doing so I had been careful to see that the battery had been disconnected. I pretended by my grimaces to be having a tremendous shock, when one of the party slipped quietly beneath the table and made the connection, and then I found my hands glued to the handles whilst I was in an agony and kicking out in all directions. Fortunately somebody soon took pity on me and I was released and I expect the boy who played me the trick had another kind of shock afterwards!' [14]

Manchester held its 'Great' exhibition in 1837, showing 'models of machinery, philosophical instruments, arts, natural history and specimens of British manufacturers'. This was well advertised with the following notice appearing in the Manchester Guardian of March 1837:

'..... The Directors beg to announce ... a POPULAR EXHIBITION of models of machinery, philosophical instruments, works in fine arts, objects in natural history and specimens of British manufacturers etc etc. In the Exhibition the Directors are desirous of affording to the working classes a convenient opportunity of inspecting the present state of our arts and manufactures and to present them with a source of rational and agreeable relaxation ...'.

These sentiments were expressed rather differently by Manchester city authorities, namely

'...to assist various Sunday-school visitors and teachers in their laudatory endeavours to amuse their scholars and to draw them from the scenes of immorality and vice incident to the race ground!' [15]

Five exhibitions were held in Manchester. The first of these, held in 1837, attracted a miscellaneous collection of items many of them curiosities or items showing skill in elaborate design and construction. Some manufacturer's techniques were exhibited, such as the manufacture of caoutchouc and India rubber articles. One such was the famous Charles Macintosh waterproof raincoat, then a Manchester speciality. Several Kammerer battery-operated electric clocks were shown and their process of electrical synchronisation demonstrated. The basement contained a display of machinery or models of machines. These included agricultural machinery, brick-making machines, several looms and sewing machines shown in operation, working steam engines and model locomotives. Several models of engines were said to be worked by 'galvanic batteries'. A major attraction to the visiting public was W.C. Day and Co.'s patent weighing machine, *'... designed for weighing ladies and gentlemen'*, with the added attraction of, *'... a proper certificate given of the particulars for the charge of one penny'*. Later exhibitions included more electrical artifacts. At the 4th exhibition in 1842–43 the electric telegraph was demonstrated by the British Telegraph Co., and electrical jars, an electrotype machine, galvanic batteries, an electromagnet with a lifting power of 1260 kg contributed by James Joule Esq, a revolving electric wheel, and a deflecting needle were exhibited. Various engines were shown, all powered by a steam engine working at the exhibition.

At their last exhibition, held in 1844–1845, Joule had added to his contribution a new galvanometer:

> *'... a coil of wire which conveys the voltaic current, is to be adjusted in the plane of the magnetic meridian. The magnetic needle in the centre of the coil will then be attracted towards a plane at right angles to its natural position and by the amount of deflection of the needle, the quantity of electricity transmitted through the coil may be measured.'*

This exhibition attracted 46 000 visitors between 18 December 1844 and 11 January 1845, showing the mechanics' institutes' exhibitions to be a major attraction in the town during the Christmas holidays. Family entertainment was again provided by an electromagnetic machine,

> *'... which continues as usual (to be), a source of very considerable merriment to hundreds of our visitors, and immediately round these machines a numerous throng gathers evening after evening, to join in the loud laughter which the shrieks and strange gestures of some tyro writhing under their effects call forth. The apparatus was fitted up by Messrs. Abraham and Dancer of this town ... it is worked by*

a small Smee's battery and ... the price is £3.13.0d ... By a very simple arrangement the intensity of the shock may be varied at the pleasure of the patient' [16].

In 1849 the Manchester Mechanics' Institute invited the British Association for the Advancement of Science to meet in Manchester and to use its rooms for their exhibition, to which it gave free entry to Institute members [17]. We can also note from the minute books of this time that the Institution in Manchester, *'... was eager to play its part in the preliminary arrangements for the Great Exhibition in London'*.

The exhibitions at Manchester were often followed with lectures on scientific subjects, often during the holiday period, as noted above. As early as 1838 the lectures for the Christmas vacation included a lecture on electricity, defined as

'... comprehending instruments for illustrating the phenomena of electricity derived from friction, galvanism, magnetism, electro-magnetism and thermo-electricity'.

A problem with most of the mechanics' institutes at this time was to obtain suitably qualified lecturers. Manchester was no exception. The lecturers were drawn from a wide field but included few men of science of the period. Faraday began a series of lectures at the Royal Institution in 1825 and was a frequent lecturer at the Adelaide Gallery, but no record exists of any connection with the mechanics' institutes. Two exceptions were John Dalton, who regularly delivered a series of lectures on atomic theory to members of the Manchester mechanics' institute, and William Staite, who demonstrated his self-regulating arc lamp to the Institute in 1850 (both were Manchester men). At Huddersfield Mechanics' Institute, the Director of the Leeds and Manchester Telegraph Offices, *'came to talk about the electric telegraph'*.

Towards the end of the 1840s the extent of the Mechanics' Institute lectures was expanded to form a series of classes on a number of subjects for youth education. During the 1840s over 1000 Manchester students attended night classes of this kind each winter. Also from 1838 onwards an alternative youth educational venture, the lyceum (the name came from America—see later), was established in Manchester. They offered a mixture of entertainment and elementary education at a lower level than the mechanics' institutes and were extremely popular, particularly so since they accommodated women for many of the lecture courses. The success of these evening lectures, arranged by both organisations, led eventually, with the demise of formal mechanics' institutes (and eventually the lyceums), to the establishment of technical schools and polytechnics in the city [3]. These changes occurred generally throughout the country and by the end of the 19th century many mechanics' institutes had become technical colleges, of which Manchester College of Technology, Glasgow Royal Technical College, Heriot-Watt College Edinburgh, Huddersfield Technical College, Leeds College of Technology and Birmingham Technical College are examples.

3.4 The Royal Victoria Gallery

Arising directly from the activities of the Manchester Mechanics' Institute was the formation of the 'Royal Victoria Gallery for the Encouragement of Practical Science'. This was modelled on the Adelaide Gallery in London, discussed in Chap. 2. Its aims were the same:

> '... to form an institution which would provide a collection of apparatus combining philosophical instruction and general entertainment; present demonstrations of elementary principles; exhibit the progress of those sciences fundamental to industry; stimulate research and invention by the award of prizes etc; and be attractive to the younger members of the Manchester community' [18].

The new Institute was proposed by Sir George Cayley, H.H. Brinley, a local Tory, William Fairbairn, Eaton Hodgkinson and John Davis of the Mechanics' Institute. Queen Victoria consented to be its patroness. Following its acceptance in 1840 the Board of Management awarded the post of first superintendent of the Gallery to William Sturgeon, who was to commence with a course of six lectures on electricity. It opened in the Manchester Exchange Dining Room and its anteroom. The collection included technical models and apparatus, dial weighing machines, electrical and philosophical apparatus, Read's mathematical instruments, a model of the electromagnetic telegraph of Wheatstone and Cooke, several electromagnets, and mechanical exhibits from Joseph Whitworth & Co. of Manchester. Lectures and demonstrations were to be planned and a scientific library established.

Sturgeon was successful in obtaining local assistance from a number of engineers and experimentalists, then becoming prominent in

Manchester scientific circles, including Telford and Wheatstone. One of these was a young man, James P. Joule, who had already made six contributions in the form of letters concerning Sturgeon's electromagnetic motor to the *Annals of Electricity*. By 1841 Sturgeon had arranged for Joule to lecture at the Royal Victoria Gallery, this lecture marking Joule's first public scientific appearance and the beginning of his scientific career.

Despite this local talent and the efforts made by Sturgeon in promoting the work of the Gallery, support for the venture did not last and the Royal Victoria Gallery closed in 1842. Sturgeon had spent the early part of his life in the Royal Artillery and whilst in the army worked assiduously at self-improvement, studying mathematics, languages and natural philosophy. This enabled him later to contribute to the work of the Adelaide Gallery before being offered the post as Superintendent to the Victoria Gallery.

The loss of his position at the Royal Victoria Gallery was a serious blow to Sturgeon and he was forced to supplement his income as an occasional lecturer at the Mechanics' Institute and the Royal Manchester Institute by becoming an itinerant lecturer in cities in the north of England in the 1840s,

> '*driving from town to town with the horse and cart which conveyed both himself and the apparatus for the illustration of his lectures*' [15].

He died in near poverty in 1850, receiving a small government pension only towards the end of his life, due to the strenuous efforts of Joule, Binney and others.

3.5 Mechanics' institutes abroad

In continental Europe, adult education had been available to the working class for some time, as special technical schools in Vienna (1875), Berlin (1876) and Antwerp (1888) and consequently the need for special institutes for mechanics was not apparent. However, mechanics' institutes did play a significant role in France. This was due to the activities of Charles Dupin, afterwards Baron Dupin and Professor of Mechanics at the Conservatoire des Arts et Metiers and a frequent visitor to the mechanics' classes given by Dr Birkbeck at Anderson's Institute in Glasgow. Dupin constantly urged that France should follow the example of the British mechanics' institutes then being formed, and by October 1825 instruction for mechanics became available in 59 French towns.

The effect of the growth of mechanics' institutes in Britain had a strong influence also on the inauguration of similar bodies in British territories abroad. Mechanics' institutes were established in Sydney and Hobart in Australia, Wellington in New Zealand, in Calcutta and Bombay in India, Canton in China and several locations in Canada.

In the United States mechanics' institutes had been in operation since 1826 when the first of these was organised in Boston. Others were established shortly afterwards in Ohio and Maryland and by the 1840s some hundreds were in operation throughout America. They established lecture courses, libraries and museums for *'the instruction of artisans and their children'* and were patronised by the leading men of science of the day. Few public exhibitions were organised in comparison with their counterparts in England. The most successful of these were the triennial exhibitions promoted by the Boston Mechanics' Institute, mainly for members and their friends. At one of these in 1878 Brush exhibited his arc-lighting system for the first time. He was successful in selling the system locally and in the process began his successful career in electrical supply which, as far as the supply of systems for arc-lighting was concerned, he was to dominate in the United States for the next two decades [19]. A few years later in 1882, a permanent Mechanics' Hall was built in Boston by the Massachusetts Charitable Mechanics' Association for the Institute and on at least one occasion used as a venue for an International Exhibition of American Products, together with arts and manufactures from foreign nations in 1883–84. This was attended by some 40 countries exhibiting nearly 680 items. It was moderately successful due largely to the recently installed electric lighting, which allowed the exhibition to remain open in the evening. It attracted about 300 000 visitors in the 18 weeks the exhibition remained open.

More successful and widespread than the American mechanics' institutes' activities were those arranged by the Franklin Institute of Philadelphia, which provided a higher grade of apprenticeship training and performed a similar function to the 'learned society' operations of the chartered engineering institutes of the United Kingdom. By 1860 mechanics' institutes in the United States had largely disappeared, many having become absorbed in Lyceums (a more successful and rival system of adult education, established by Josiah Holbrook, which flourished until shortly after the civil war) and other educational establishments. But the Franklin Institute has remained an important research and teaching centre and highly influential in promoting technological exhibitions.

3.6 The Franklin Institute

The Franklin Institute of the State of Pennsylvania in the United States was established in 1824 to provide technical and vocational training in science and technology. It received much help and support from the staff of the University of Pennsylvania and its lecture programmes were of significant quality, with visiting lecturers such as Dr F.W. Aston and Sir J.J.Thompson from Britain in 1922–23. At the time that the mechanics' institutes of Great Britain were promoting technical exhibitions the Franklin Institute was also busy setting up a series of technical exhibitions for its members and the general public [20]. Some 26 exhibitions were held between 1824 and 1858, with a 27th in 1874. The exhibition of 1858, the last one to be held before the civil war, attracted 100 000 visitors and, like the previous exhibitions, was billed as '...*an exhibition of products of the American industry*'. After 1858 this view was changed and the Institute resolved to '...*support industry, not by a competition of local advantage, but by a competition of intellect*'. This was seen in the 1874 Franklin Institute Exhibition, arranged to celebrate the 50th anniversary of the Institute. It attracted 1200 exhibitors showing a range of manufactures in the engineering and technological fields. Some 268 000 visitors attended this exhibition, considered by the Institute's organisers as being, '.. *a rehearsal for the great Centennial Exhibition to be held two years later, in 1876*' [21].

It also helped to prepare the way for two major international exhibitions, also held in Philadelphia: the International Electrical Exhibition in 1884 and the 1920 Sesqui-Centennial Exposition (both of which, together with the 1876 Centennial Exhibition, are considered later). The work of the Franklin Institute in these government-supported exhibitions formed part of the help given to the administration of the United States for its participation in overseas international exhibitions, particularly those held in Paris, Vienna and Frankfurt in the latter half of the 19th century. Two important events in the United States' electrical engineering history took place at the time of the 1884 International Exhibition which are worthy of note. The United States' government, supportive of the work of the Franklin Institute since its inception, proposed that a National Conference of Electricians be held at the Franklin Institute, ' ... *during a period when so many men interested in electrical science would naturally gather at*

Philadelphia' [22]. Additionally on the occasion of this conference the Franklin Institute was instrumental in founding the American Institute of Electrical Engineers, predecessor of the Institute of Electrical and Electronics Engineers (IEEE). Two further exhibitions, both national in character, were organised in Philadelphia by the Institute before the close of the century. These were the Exhibition of Novelties of American Manufacture in 1887, in which 375 manufacturers contributed their products for display, and a National Export Exhibition in 1899, both events we would now designate as trade exhibitions.

As part of its 'learned society' activities the *Journal of the Franklin Institute* was established in 1826. It is still in publication today and regularly carries accounts of new devices, processes and techniques in science and engineering. It receives many papers from the international community in science and technology. Michael Faraday was a frequent contributor in the 1840s [23].

3.7 British Association for the Advancement of Science

The British Association for the Advancement of Science (BAAS), mentioned earlier in connection with its visit to the Manchester Mechanics' Institute, was founded in 1831 as a venture in scientific public relations to satisfy the wide interest of the people at the time in scientific matters (although in practice the audience at the BAAS meetings were generally the articulate middle class of Victorian society and fellow scientists). Its role has changed but slightly today and continues in a determined effort to make the public more aware of the role of science and technology in their daily life. It has always had the support of the men of science of the day and enrolled their help in a programme of lectures at their yearly meetings and, at some of its earlier meetings, supported exhibitions on science and technology. It was not always so successful however in enlisting the support of the Victorian engineers and technologists. Indeed they were not particularly welcome by some members at the BAAS meetings. The Secretary (Murchison) in a letter to the President in 1838 commented that, *'...Boulton and Watt, the Gog and Magog of the new era, have cut all connection with the place and take no part in its doings'* [24].

Their meetings are not held at any fixed location and the organisers, *'... engage in a ceaseless peregrination round the cities of the British Isles and later, of the British Empire'* [25]. Their hosts at these locations were sometimes successful in coinciding their invitations to the BAAS with technological exhibitions. Some examples were given in this and the previous Chapter, with meetings arranged in Manchester to coincide with the Jubilee exhibition of 1887 and in Birmingham in 1839 and 1849 at the venue for National Exhibitions at Bingley Hall. One hundred years later in 1949 the British Association were back in Birmingham, this time at the National Exhibition Centre, where the 1949 National Radio and Television Exhibition was taking place. The BAAS also played a significant part with the organisers of the 1923–1924 Empire Exhibition at Wembley.

Demonstrations of scientific or technical matters were often given at the BAAS meetings at one or other of the section meetings. In 1833 at the Cambridge meeting Saxton was invited to show his electro-magnetic engine at a time when he was busy demonstrating this at the Adelaide Gallery. The Glasgow meeting in 1876 saw the earliest example of the Bell Telephone to reach Britain, brought back to this country by Sir William Preece, and in 1881 Ayrton and Perry exhibited their ammeter and power meter for the first time at the BAAS meeting in York [26].

Oliver Lodge was to present many of his ideas in papers presented at meetings of the BAAS; he attended well over 50 annual meetings. The meeting at Dover in September 1849 coincided with the centenary of Volta's discovery of electric current, and Sir Ambrose Fleming gave a lecture to the BAAS on *The centenary of the electric current*. At this lecture Marconi was invited to give a demonstration of wireless telegraphy, transmitting from an aerial, mounted on the top of the town hall at Dover, to Goodwin Sands lightship. At the Oxford meeting in 1894 Lodge also gave a public demonstration of transmitting information by radio using Morse code during a lecture on the work of Heinrich Hertz [27].

Initially the BAAS arranged a number of its own exhibitions of

'... *manufacturers and inventions, beginning with that at Newcastle in 1838. These exhibitions of British private enterprise were important in places like Newcastle (1838), Birmingham (1839), Glasgow (1840) and Manchester (1842), and served to provide publicity, information and practical criticism for manufacturers and merchants alike*' [28].

Not all manufacturers welcomed the publicity attendant on these visits and exhibitions. James Watt commented, '... *It was widely feared that industrial exhibitions would reveal trade secrets and thus reduce Britain's technological advantage*'. He directed his comments also to the mechanics' institutes' exhibitions since the BAAS and the mechanics' institutes were the chief promoters of provincial exhibitions at that time.

The BAAS Exhibition at York in 1831 (its first meeting) exhibited a number of applications of magnetism, including one by Scoresby, '... *to measure the thickness of rocks (otherwise immeasurable) by the action of a pair of bar magnets and a compass needle*'. A member named Potter was willing, '... *to contribute on some phenomena of electrodynamics, exhibited in experiments, for which he needed an electrical machine, a Leiden jar and a galvanic trough, from the scientific gentlemen of York*'. Abraham (from Sheffield) was exhibiting an apparatus, '.. *illustrating some recent discoveries connected with magnetism*' [29].

It was at this inaugural meeting that Charles Babbage, following a suggestion that members could bring some of their portable experimental apparatus to the meetings, formally proposed that

> '... *might it not be possible to have an exhibition of manufacturers at each meeting? I am induced to insist on this ... from the experience I have had of the great utility ... of carrying such portable samples of art or nature as they can convey with them*'.

This was not to be realised, however, and only three major exhibitions were mounted: those at York (1831), Newcastle (1838) and Glasgow (1840). The Newcastle Exhibition was attended by 3000 visitors and was remarkable only for a demonstration of Hawthorn's steam engine. The Glasgow Exhibition was also to show a noisy exhibit, that of the Perkin's steam gun (last seen at the Adelaide Gallery in London).

In 1848 the American Association for the Advancement of Science (AAAS) was formed in Philadelphia, modelled on the BAAS, and arranged its Pacific Coast meeting in 1915 to coincide with the San Francisco Exposition (see Chap. 10).

Forty years later the BAAS extended its circulating meetings to include various locations within the British Empire. These were arranged at a number of venues: Montreal (1884), Toronto (1897), South Africa (1905), Winnipeg (1909) and Australia (1914). At the Toronto meeting William Duddell's galvanometer (termed by him an oscillograph) was shown for the first time by its inventor [30].

No meetings were arranged during the war years and afterwards, in 1919, the BAAS modified its role to being a forum for discussion and analysis of problems which have a social and cultural implication;

although this attracted the attention of the media, thus ensuring wide publicity, it resulted in scientists and technologists ceasing to use the occasion of the yearly meetings to announce new discoveries [25].

3.8 References

1 BERLYN, P.: 'A popular narrative of the origin, history, progress and prospects of the Great Industrial Exhibition 1851' (James Gilbert, London, 1851)
2 BRIGGS, A.: 'Victorian people' (Penguin Books, London, 1970), p. 128
3 SINGER, C., HOLMYARD, E.S., HALL, A.R., and WILLIAMS, T.I.: 'A history of science, technology and philosophy, Vol. V—1850–1900' (Oxford University Press, 1958), Chap. 32
4 KELLY, T.: 'George Birkbeck, Pioneer of Adult Education' (University Press, Liverpool, 1957)
5 'Rules and orders of the Mechanics Institute, 1823', *The Mechanics Weekly Journal,* 15 Nov. 1823
6 *Mechanics' Institute Exhibition Gazette,* 27 June 1840, pp. 46–47
7 KUSAMITSU TOSHIO: 'Great exhibitions before 1851', *History Workshop Journal,* Spring 1980 (9), pp. 71–89
8 *Leicester Journal,* 28 August 1840
9 Secretary to the Manchester School. Letter to 'The Working Men of Manchester', *Manchester Guardian,* 8 January 1845; also *The Exhibition Gazette,* Manchester Mechanics' Institute, 25 January 1845, p. 40
10 HADFIELD, C.: 'Atmospheric Railways' (David and Charles, Publishers, Newton Abbot, 1967)
11 'Working models of the atmospheric railway', *The Exhibition Gazette,* Manchester Mechanics Institution, No. 6, 25 January 1845, p. 71
12 GALLAGHER, M., and DE PAOR, A.: 'A mere plaything, or the noblest medicine yet?' 21st IEE History of Technology Weekend, July 1993, pp. 77–122
13 HEYWOOD, SIR BENJAMIN: 'Addresses delivered at the Manchester Mechanics' Institute (1825–1840)', *North of England Magazine,* 1843, **3** (20), pp. 263–266
14 HAYES, L.M.: 'Reminiscences of Manchester' (Sherratt and Hughes, Manchester, 1905, pp. 25–27
15 TYLECOTE, M.: 'The Mechanics' Institutes of Lancashire and Yorkshire before 1851' (Manchester University Press, 1951), p. 183
16 'The electro-magnetic machines', *The Exhibition Gazette* (Manchester Mechanics Institution, 1845), p. 55
17 'Manchester Mechanics' Institution Minute Book', 2 August 1849; also see MMIMB for 21 July 1841 and 2 June 1842
18 KARGAN, R.H.: 'Science in Victorian Manchester' (Manchester University Press, 1977), pp. 36–41
19 PASSER, H.C.: 'The Electrical Manufacturers 1875–1900' (Harvard University Press, 1953)
20 SINCLAIR, B. 'Philadelphia's Philosopher Mechanics' (Johns Hopkins University Press, Philadelphia, 1924)
21 COULSON, T.: 'The Franklin Institute from 1824 to 1949', *J. Franklin Inst.,* 1950, **249** (1), p. 18
22 POMERANTZ, M.A. (Ed.): 'Electrical Conference at Philadelphia', *J. Franklin Inst.,* 1966, **301**, pp. 4–5
23 OLIVER, J.W.: 'History of American Technology' (Ronald Press Co., New York, 1956)
24 MACLEOD, R, and COLLINS, P. (Eds.): 'Early correspondence of the BAAS' (Science Review Ltd., 1981)
25 MORRELL, J., and THACKRAY, A.S.: 'Gentlemen of Science' (Oxford University Press, 1981)

26 '*The Electrician*', 1886, **9**, pp. 375, 413
27 ROWLANDS, P., and WILSON, J.P.: 'Oliver Lodge and the invention of radio' (P.D. Publications, Liverpool, 1994)
28 'Catalogue of the illustration and manufactures, inventions and models, philosophical instruments of the 2nd Exhibition of the BAAS Birmingham meeting'. BAAS archives, 1839
29 'Report of the BAAS', BAAS archives, 1831, Nos. 77, 88 and 94
30 '*The Electrician*', 1897, **39**, p. 637

Chapter 4
The Royal Society of Arts

4.1 The engineer in Victorian society

The Victorian age, which corresponded so closely with the Industrial Revolution, was suffused with optimism and self-confidence. Changes in public life at that time were startling and seen by most people as beneficial. Roads were improved and extended, an extensive canal system was in place, railways were beginning to reach all the main centres of communication and, in the middle of this period, the boon of telegraphy was becoming apparent to all. A vast building programme was under way and impressive structures were seen rising in all the major towns, rivers were being spanned by huge iron bridges to allow the railways to expand, and iron ships, powered by steam, were transporting goods and people to all quarters of a successful and expanding Empire. The Victorian people had no doubt who were responsible for these widespread developments and the accompanying wealth they created. Engineers and men of science were respected, known by name, and their deeds followed with enthusiasm. In the 1870s the headmaster of a large public school was recorded as saying that, '... *three out of every four of his pupils would, if polled, declare for engineering*' [1]. They were the heroes of the age and served as what we now know as 'role models' to the population at large:

> *'When Robert Stephenson died in 1859,' writes Rolt, 'the whole nation mourned him. By special permission of the Queen, his funeral cortege passed through Hyde Park on its way to Westminster Abbey, where his body was buried beside Thomas Telford and the whole route lined with silent crowds.... In Newcastle the 1,500 employees of Robert Stephenson and Company marched through*

silent streets to a memorial service. It was as though a King had died.... Never again would a British Engineer command so much esteem and affection; never again would the profession stand so high' [2].

Table 4.1 illustrates that many of the scientists and engineers of the day were less than 50 years of age in 1851. They were highly productive and innovative and intensely curious about the possibilities of electrical power. William Armstrong, later Lord Armstrong, was one of the leading industrialists of the Victorian era. In 1880 his home at Cragside, near Rotherbury, was the first house fitted with electric lamps in Britain. Other industrialists and scientists were adding to

Table 4.1   Scientists and engineers—age in 1851

Carl Gauss	74	George Corliss	34
Hans Oersted	74	James Joule	33
William Cubitt	66	Alexander Bain	33
Francois Arago	65	Cromwell Varley	32
George Ohm	64	Hermann von Helmholtz	30
Robert Stirling	61	Josiah Clark	29
Samuel Morse	60	William Siemens	28
Michael Faraday	60	William Thomson	27
Charles Babbage	59	Gustave Kirchhoff	27
Joseph Henry	54	Zénobe Gramme	25
Charles Wheatstone	49	Joseph Swan	23
Daniel Rühmkorff	48	Moritz Jacobi	23
Robert Stephenson	48	Joseph Saxton	22
Joseph Whitworth	48	Charles Bright	19
Fothergill Cooke	47	David Hughes	19
Wilhelm Weber	47	Gustave Eiffel	19
Louis Breguet	47	Fleeming Jenkin	18
John Brett	46	William Preece	17
Isambard Brunel	45	Marcel Deprez	8
Henry Cole	43	Rookes Crompton	6
William Saite	42	George Westinghouse	5
William Armstrong	41	Thomas Edison	4
William Henley	38	William Ayrton	4
Henry Bessemer	38	Alexander Graham Bell	4
Werner von Siemens	35	John Hopkinson	2
		Edward Weston	1

William Sturgeon had been dead for 1 year, George Stephenson for 3 years, John Daniell for 6 years, George Birkbeck for 10 years, and André Ampère for 15 years.
Charles Parsons, Nikola Tesla and Sebastian Ferranti had yet to be born

their original activities experimentation and application of electrical power. Hermann Helmholtz, initially trained in medicine, transferred his attention to physics, and in 1858 to experimental science and electricity. Samuel Morse achieved considerable success as a portrait painter before taking a lifelong interest in electrical communication. Charles Wheatstone accompanied his successful early researches into acoustics with the development of telegraph equipment. Joseph Swan followed his work in chemistry with the development of incandescent lamps. Michael Faraday, the most striking example of them all, broadened his work as an analytical chemist to embrace the investigation of electricity and magnetism with seminal results for the history of electrical science. All of these, together with the stalwarts of the mechanical age—Telford, Stephenson, Brunel, Arkwright and many others—would be known to the Victorian public through their constructions and inventions.

Less well known to the public but just as respected were the various associations of engineers and technologists which had arisen to serve their needs with the growth of trade and industry. The earliest of these, dating from the 18th century, were the Royal Society of Dublin and the London Society of Arts. Proposals for the formation of the Society of Arts included the words, '... *for the preservation of Operative Knowledge, Mechanical Arts, Inventions and Manufactures*', thus clearly indicating the earlier meaning of the term 'arts', so different from its modern usage. This London Society became the precursor to the 'Society for the Encouragement of Arts, Manufactures and Commerce', mentioned in Chapter 2. It was generally known, however as, 'The Society of Arts', to become in 1847 'The Royal Society of Arts' (RSA), following the award of a Royal Charter from the Queen in that year. Somewhat earlier it became established in a magnificent set of rooms in John Adam Street, London, the imposing entrance of which is illustrated in Figure 4.1.

4.2 The Society of Arts

The Society was established in 1754, and from its beginnings, meeting in Rawthmell's coffee house in Covent Garden, was a strong supporter of industry and the industrial worker. Although it never established a strong role in formal education for the technologist, as did the mechanics' institutes, it did provide encouragement in the form of rewards for technical expertise. They consisted of prizes for '...*promoting of improvements in the liberal arts and sciences, manufactures*

Figure 4.1   The Royal Society of Arts in John Adam Street
(Wood, 'A History of the Royal Society of Arts', 1913)

etc.' and took the form of money prizes, premiums and medals. The Society took considerable pains to organise these so as to support their various objectives, which invariably meant the encouragement of effort by the participants in one of the arts or industries favoured by a number of committees set up for this purpose [3].

Mention must also be made here of similar efforts being made in Scotland by the 'Select Society of Edinburgh', formed in 1754, to hold exhibitions and award '... *premiums for the discovery of any Useful Inventions in Art and Science'*. Exhibitions were held yearly between 1755 and 1786, when the work of the Edinburgh society came to an end. Much later the Royal Scottish Society of Arts was to carry on this role, as described later in this Chapter.

4.3 The exhibitions

Participants of the RSA competitions, leading to the award of premiums, were encouraged to submit practical evidence of their accomplishments, and this invariably resulted in models or demonstration apparatus being left with the Society. The idea of using these to form an exhibition in the Society's rooms for the benefit of members and their guests was due to a life-long member and sometime Secretary of the Society, Francis Whishaw, an engineer who was later to become immersed in the development of the electric telegraph. It quickly became an established practice to gather these submissions together from time to time and exhibit them in the Society's rooms (Figure 4.2). Lists of these items were prepared and, in effect, took the form of an exhibition catalogue. The list for 1773, for example, prepared by A.M.Bailey, Registrar for the Society and son of its first curator, is illustrated in two volumes and contains a fine and detailed record of the sort of entries submitted by the technologists of the time [4].

The first such exhibition was held in 1761, with subsequent exhibitions arranged intermittently until 1861. This first exhibition was devoted entirely to models of mechanical structures. The curator, William Bailey, '*a man of considerable mechanical knowledge'*, was engaged to take care of these and other models, submitted later. They included models of windmills, spinning wheels, a crane, a threshing machine and several scale models of ships. Later exhibitions became somewhat wider in scope and included inventions and experimental equipment. To some extent the artifacts submitted were limited by the Society's rules, which excluded all articles subject to patent protection. However, this does not seem to have affected the electrical items shown in later exhibitions.

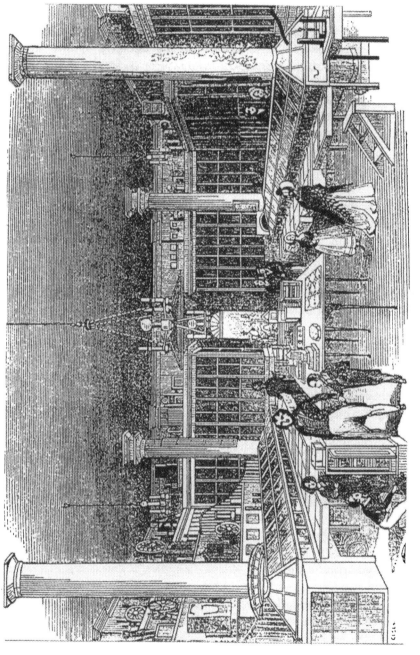

Figure 4.2   The Royal Society of Arts Repository in 1773 (Knight's 'London', 1843)

4.4 Some technical exhibits

Initially the exhibits at the Society's exhibitions were drawn mainly from the fields of agriculture, mechanical engineering and chemistry, but from early in the 19th century the Society began to award premiums and exhibit items of interest to communications and electrical specialists. One of the first of these was a gold medal to Admiral Popham in 1816 for his mechanical telegraph—a semaphore arrangement in which two masts were employed, each with an arm capable of being set at any desired angle to the vertical. This apparatus was used between the Admiralty, Portsmouth, other naval stations and ships at sea for a number of years until it was superseded by the electric telegraph.

The first definite plan for an electromagnetic telegraph was shown at an exhibition at the Royal Scottish Society of Arts in Edinburgh in November 1837. This was William Alexander's Electric Telegraph shown in Figure 4.3. It was a needle telegraph consisting of 30 magnetic needles, each carrying a light screen which is rotated out of view when affected by a current-carrying conductor surrounding the needle. In doing so a printed letter is revealed as shown in the diagram. Thirty copper wires and a common return wire connect the transmitted keyboard with a panel of such indicators. Professor Alexander initially tested the device over four miles of connecting wire in a chemistry classroom at the University of Edinburgh. It was never used to provide a commercial service, however, being overtaken by other, less elaborate, devices, later to be seen at the 1851 Great Exhibition in London

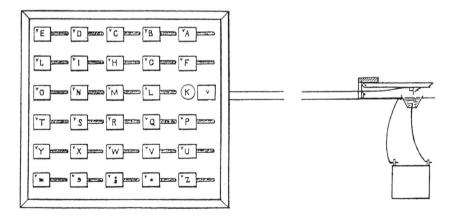

Figure 4.3   Alexander's Needle Electric Telegraph
(Official Catalogue of the Great Exhibition, 1851)

At a later Royal Scottish Society of Arts exhibition in 1842, billed as The Electro-magnetic Exhibition, Robert Davidson, a Scotsman, showed models of his electric locomotive, a full-sized version of which had been financed by the Society and driven on the Edinburgh and Glasgow Railway. This was a remarkably early example of the application of electric power to vehicle design. The four-wheeled carriage he used (shown in Figure 4.4) was 4.9 m in length and weighed ~5 t. The simple rotary motors he used consisted of wooden cylinders on each axle with iron strips fixed in grooves in the cylindrical surface. Horseshoe magnets—one either side of the cylinder—were energised alternately through a simple commutator carried on the axles, and produced a succession of magnetic pulls on the iron bars which caused the carriage axles to rotate. Zinc-acid batteries were carried and arranged at each end of the carriage. The plates of the cells could be raised out of wooden troughs containing the electrolyte by a windlass arrangement to cut off the supply of current as required. The carriage ran at ~6.5 km/h on a level track, with the motors having an efficiency of ~10% [5]. The demise of this invention came on a return trip from Aberdeen, where an exhibition had been arranged, when the electric carriage came to a mysterious end and it was suspected that steam engineers were responsible for the damage in fear that they would be deprived of their livelihood by the new invention [6]. Other electrical devices shown at this exhibition

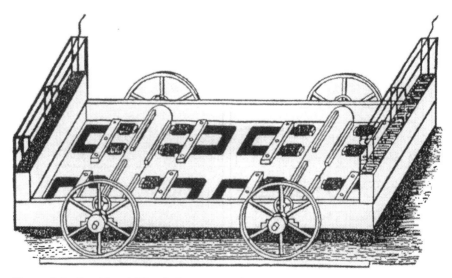

Figure 4.4   Davidson's Electric Locomotive shown at the Royal Scottish Society of Arts in 1842
(Bowers, 'History of Electric Light & Power', 1982)

ELECTRO-MAGNETIC
EXHIBITION,
UNDER THE PATRONAGE OF THE
Royal Scottish Society of Arts.

Mr. ROBERT DAVIDSON'S
EXHIBITION OF ELECTRO-MAGNETISM,
AS A MOVING POWER,
IS NOW OPEN!
IN THE
EGYPTIAN HALL, *Piccadilly.*

THE MODELS AND APPARATUS COMPRISE THE FOLLOWING:

A LOCOMOTIVE ENGINE,
Carrying Passengers on a Circular Railway.

A PRINTING MACHINE AND TURNING LATHE.
A SAW MILL.
A MACHINE for COMMUNICATING the ELECTRO-MAGNETIC SHOCK.

AN ELECTRO-MAGNET!
The largest ever made. It weighs upwards of 500 Pounds, and will suspend many Tons.

A GALVANIC TELEGRAPH.
THE COMBUSTION OF METALS,
Attended in each case with Splendid Coruscations, peculiar in color to the Metal operated upon, &c &c.

MODEL OF THE FLYING MACHINE

Balloon Navigation Illustrated.

13,000 small Nails formed into a Rope)
MODEL OF AN ÆRIAL CARRIAGE,
By SIR GEORGE CAYLEY.

OPEN FROM 12 TILL 6, AND FROM 7 TILL 9.
Admission, ONE SHILLING.
Children under 12 Years of Age, SIXPENCE.

PRINTED BY ELECTRO-MAGNETISM.

Figure 4.5   Poster advertising an exhibition of Davidson's machines
(Bowers, 'History of Electric Light & Power, 1982)

included a printing machine and a lathe, both operated by electricity, an electromagnetic telegraph and various mechanical models as described in a poster advertising this event (Figure 4.5).

The Society of Arts was particularly rich in its acquisition of telegraphy-related items. In 1860 Professor Hughes displayed his printing telegraph in the Society's rooms. It consisted of a set of keys representing the letters of the alphabet, arranged as in the keyboard of a piano, which were interconnected with a complex clock mechanism, controlled by electromagnets. When a given key was depressed a current flowed through an electromagnet, starting a chain of movements through the mechanism so that the setting of the instrument corresponded to that particular letter. At the same time the current flowed to an instrument at the receiving station to produce the same settings in its clockwork device corresponding to the letter, which was then printed.

Many examples of telegraph cables were shown in the Society's rooms, using India rubber and gutta-percha as the insulating medium. Gutta-percha, the gum of the percha tree (Malay: getah percha) was first revealed through a detailed account of its properties sent to the Society of Arts in 1843 by William Montgomerie, a resident of Singapore. It was subsequently used for the first time in insulating the underground cables laid between Dover and Folkestone in 1849.

Cromwell Varley was to make the first public exhibition of his electrometer in the Society's rooms and Robert Walker contributed his platinised graphite battery, which he claimed, '... *once charged would continue in good working order for six to twelve months without any attention'* [7]. Also rewarded with a Society prize, in 1840, was Alfred Smee's galvanic battery, which was exhibited as a gold medal award.

Of major interest to electrical engineers is the apparatus of William Sturgeon, who submitted his invention of the first practical soft-iron electromagnet to the Society in 1825 which was also duly exhibited in the Society's rooms [8] and is shown in Figure 4.6. This was awarded a silver medal and a prize of 30 guineas. Electrodeposition was another subject rewarded with a silver medal in 1841. This was to Robert Murray for his method of obtaining a conductive surface for electrodeposition by means of plumbago, a method still in general use [9].

Some bizarre uses for electricity were also displayed. One such, submitted for the 9th exhibition in 1857, was by J.F. Caplin, MD, of London and described as an electrochemical bath, the object of which was to remove from the (patient's) system any particles of metal which may have been absorbed by it. He declared its mode of use as follows:

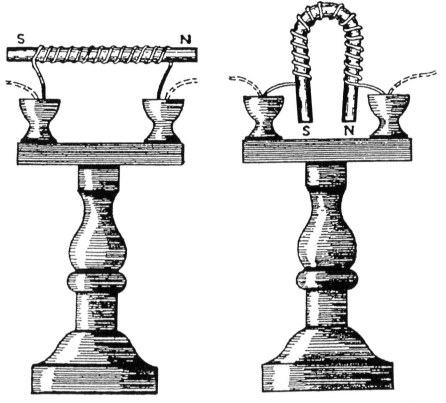

Figure 4.6   Sturgeon's soft-iron electromagnet exhibited at the Royal Society of Arts, 1825
(Wood, 'A History of the Royal Society of Arts', 1913)

'...*The patient is placed up to the neck in a metallic bathing-tub, isolated from the ground, and made to rest in a horizontal position upon a wooden bench .... The water is to be acidulated with nitric or hydrochloric acid for the removal of mercury, gold and silver, and with sulphuric acid for the removal of lead. One extremity of the bath is put in contact with the negative pole of the pile, ... and the patient takes hold of the positive pole, sometimes with the right hand and sometimes with the left. The arm is held up by supports in contact with the seat. The extremity of the positive conductor, which the patient holds, is armed with a massive iron handle, wrapped around with linen, to diminish the calorific action of the current, which is very powerful, and which, without this prevention, would burn the hands. ...the positive current enters by the arm, circulated from the head to the feet, and is neutralised at the negative pole on the sides of the bathing tub. ... the electric fluid radiates from the*

> *body into the bath, forming a multitude of currents from the entire surface of the body which, after having traversed the organs and even the bones, neutralise themselves upon the negative side of the bathing tub'* [7].

There appears to be no record of whether this drastic treatment was actually tried on a patient!

4.5 Award of medals

The various classes of medals awarded to exhibitors was a matter of prime importance to the industrial life of the Victorian era, and indeed for almost a century afterwards a similar system has been followed at almost all major exhibitions. The medals awarded were much more than a matter of pride to the successful entrant. They were a sound source of publicity, an aid to profits in an age of commercial competition and did much to stimulate emulation and diffuse knowledge about a new process or technique.

The Society of Arts may be said to have initiated this practice of medal awards at public exhibitions, which reached such prime importance at the 1851 Great Exhibition. The issue of medals to indicate academic or industrial prowess to individuals also started with the Society. Their premier medal, the Albert Medal, first awarded in 1864 for *'distinguished merit in promoting arts, manufactures and commerce'*, is awarded annually and has been given to a number of electrical pioneers, commencing with Michael Faraday in 1866 and including Fothergill Cooke and Charles Wheatstone (1867), William Siemens (1874), George Airy (1876), Lord Kelvin (1879), James Prescott Joule (1880), Louis Helmholtz (1888), Thomas Edison (1892), David Hughes (1896), Henry Wilde (1900), Graham Bell (1902), Lord Rayleigh (1905), Joseph Swan (1906) and Charles Parsons (1911).

4.6 Proposal for a national exhibition

The activity of the Society in the public exhibition, known as 'The National Repository', was noted in Chapter 2. Although not entirely successful, this experience, together with that of the smaller exhibitions held in the Society's rooms and the success of the French trade expositions, led Francis Whishaw, then Secretary to the Society, to consider establishing a 'Grand Annual Exhibition of

Manufacturers'. As a precursor to this aim he offered, from his own purse, prizes amounting to a total of £300 for various types of paintings and 'useful inventions', and with these he set up in 1844 a single evening's exhibition of pictures and mechanical inventions. This was repeated a month later and caused much discussion amongst members and their friends on the possibility of a much larger exhibition. Amongst those interested was William F. Cooke, who made an outline proposal for such an exhibition. He did this (probably by intent) just before the visit of Prince Albert, who as President of the Society was to present the prizes. Whishaw made good use of this opportunity and, although five years were to pass before the idea was acted upon, the meeting provided the seed which was eventually to lead to the 'Great International Exhibition of the Work of all Nations' in 1851 described in the following Chapter.

4.7 Arts Society exhibitions after 1851

The Society was active in arranging exhibitions for a considerable number of years, either for its members and their guests in its own premises, or assisting outside bodies, such as the National Repository, and culminating in its efforts on behalf of the 1851 and 1862 Great Exhibitions in London. After the 1851 exhibition the Society continued to hold small exhibitions in its own rooms at the Adelphi near Charing Cross. It had already held an 'Exhibition of Recent Inventions' in 1848, which contained 446 items. This was the first of a yearly series of 14 'inventions' exhibitions which continued up to the year of the second Great Exhibition in London in 1862 when the series terminated.

In 1854 the Society of Arts supported an educational exhibition of some technical interest. It was hoped at the time that this would '... *do for Education what the 1851 exhibition had done for Manufacturers'* and was to some extent a response to Britain's failure in the technical educational field [10]. Some of the exhibits transferred from the South Kensington Museum may be seen in Figure 4.7. These included a large collection of apparatus for, *'galvanic, voltaic and frictional electricity'*, shown within several glass cases [11]. Overhead the gallery contained a rich display of scientific apparatus with mechanical engineering well represented. This included a machine illustrating centrifugal forces, sectional models of steam engines and apparatus for teaching 'natural philosophy', which turned out to be exhibits in natural history and astronomy.

Finally, reaching almost the status of an exhibition in its own right for the interest shown by its members was the facility installed by the Society in 1882 to provide electric light in all its rooms. A Siemens dynamo was installed in a cellar, driven by a gas engine, and as an insurance against failure a storage battery was added in 1885. In 1889 electricity was obtained from the newly established street mains.

*Figure 4.7   Exhibits at the Educational Exhibition, London, 1854
(J. RSA, 1854, 2)*

4.8 References

1 ESCOTT, T.H.S.: 'England, its people, politics and pursuits' (Chapman and Hall, London, 1885), pp. 554–55
2 ROLT, L.T.C.: 'Victorian engineering' (Allen Lane, London, 1970), p. 161
3 WOOD, H.T.: 'A history of the Royal Society of Arts' (John Murray, London, 1913)
4 BAILEY, A.M.: 'Description of the ... machines... in the Repository of the Society.' 2 Vols, RSA Archives, London, 1773
5 BOWERS, B.: 'A history of electric light and power' (Peter Peregrinus, London, 1982), p. 50
6 'Davidson's motors', *Philos. Mag.*, 1839, **15**, p. 350; also in *The Electrician*, 1882, **9**, p. 400
7 'Catalogue of the 7th to 13th exhibitions of inventions'. Report for RSA (Trounce, London, 1855–1861)
8 *Trans. RSA*, 1825, **XLIII**, p. 37
9 *Trans. RSA*, 1841, **LIII**, Pt. 2, p. 10
10 'Educational exhibition at the Royal Society of Arts', *J. RSA*, 1972, **120**, pp. 183, 253
11 'Educational exhibition of 1854', *J. RSA*, 1854, **2**, pp. 588–589

Chapter 5
The Great Exhibition of 1851

5.1 Early ideas on international exhibitions

In Europe after about 1840, the influence of earlier national exhibitions together with the success of the mechanics' institutes and Royal Society of Arts exhibitions caused many influential people in both Britain and France to consider a comprehensive exhibition which would, unlike previous exhibitions, be international in character, gathering together the competing industrial products of many nations. France was the first to voice this idea through the office of the President of the Société Royale d'Emulation, M. Boucher de Perthes. He called for an 'exposition universelle' in 1834. This came to nothing, however, and it was Great Britain through the enterprise of the Royal Society of Arts and Prince Albert which finally brought this into effect. Mindful of the failure of an earlier attempt to propose such an exhibition by William Cooke in 1844, a venture considered by many to be both financially and physically hazardous (riots by the visiting populace to London were feared!), the Society once again decided to test the feeling for a large technological exhibition in London and to this end set up a range of prizes to be awarded for items submitted to a proposed series of annual 'exhibitions of select specimens of British manufacture and decorative art', the first of these to be held in 1846. Three more annual exhibitions were funded, in 1847, 1848 and 1849, each more successful than the last in encouraging both exhibitors and visitors, with the final exhibition admitting over 100 000 visitors before it closed its doors after seven weeks.

5.2 The Great Exhibition

With the successful results of these exhibitions the Society now felt confident that the time was ripe to take up the offer made by the Prince Consort in 1844 that '... *as soon as the plan for carrying it (the proposed exhibition) into effect should be matured, to lay it before His Royal Highness'* [1]. At a meeting in Buckingham Palace in June 1849, Prince Albert agreed to act as President for a proposed Royal Commission which was to determine the way in which a *'great exhibition of the works of all nations'* should be arranged. He himself suggested four great divisions of: raw material; machinery and mechanical inventions; manufacturers; and sculpture and plastic arts, each with its own group of advisers and organisers. Broadly this structure was accepted but more divisions became necessary and the whole subject of classification was thoroughly discussed at several meetings of the organising committees (see later).

The Great Exhibition was to be an international exhibition on a grand scale with all manufacturing countries represented. Over 6000 foreign exhibitors submitted their products and over 7000 from the United Kingdom. The exhibition remained open for 146 days and was seen by a staggering total of 6 040 000 visitors[1]. At its peak it was admitting some 40 000 visitors per day spread around the glass-covered site. This became known by its popular title of 'The Crystal Palace', and remains a unique architectural and engineering achievement of the Victorian era [2–4].

5.3 The Crystal Palace

Remarkably its conception was the work of an erstwhile head gardener, Joseph (later Sir Joseph) Paxton, who modelled the design on his large greenhouses, recently erected in Chatsworth Gardens. Paxton was no mere gardener, however. At the time of his design work on the Crystal Palace he was a rich man and a Director of the Midland Railway Co., and was said to display a talent, *'comparable with that of Capability Brown in an earlier generation'*.

A competition for a suitable design to be erected in Hyde Park, the chosen venue for the Exhibition, was announced in 1850 and the results displayed to the public in the rooms of the Institution of Mechanical Engineers. None of the 233 designs and ideas submitted

---

[1]The influx of millions of visitors to a major international exhibition after 1851 became the norm, as may be seen from Table 2.1, and which continues to the present day.

fulfilled the requirements of ease of assembly (and disassembly—its stay in Hyde Park was strictly limited to the duration of the exhibition), large uncluttered enclosed space for the exhibits, good use of natural daylight, and cheapness of construction. Paxton's design, which was submitted later, was to use a cast-iron and wrought-iron framework and an enormous number of glass panels. Glass was one of the newly freed commodities from which a prohibitive tax had just been removed and Paxton took full advantage of this. His design had no difficulty in fulfilling the specification requirements as well as containing an extraordinary number of new and ingenious features to shorten the work of erection. His plans for the building would be designated a 'feasibility study' today and the actual design calculations and working drawings were made by Fox, Henderson Company of Smethwick. The basic components were girder sections, similar to those used since 1840 on the many bridges required for the new railways, cast-iron hollow columns which additionally served to convey rain-water collected by the triangular roof sections, and standardised glass panels of the largest size then possible to manufacture. Similar parts were interchangeable and mass-produced—a new technique for the building industry [5]. A contemporary drawing of the great central transept is shown in Figure 5.1. The total area enclosed was 7.39 ha and included within the building a row of huge elm trees, thus solving a conservation problem which had defeated other submitted designs. The first column was raised on 26 September 1850. Within 17 weeks, nearly a million feet of glass had been attached to the web-like iron structure—a remarkable achievement by any standards! To achieve this speed of erection some 2000 men were employed aided by four powerful steam engines and a stable of horses. A number of power-operated machining tools were used on the site (Figure 5.2). The principal organiser responsible for the work of construction was a civil engineer, Henry (later Sir Henry) Cole, who commissioned the building, and Sir William Cubitt, President of the Institution of Civil Engineers, who superintended its construction. Full information on this project, probably the most outstanding architectural achievement of the century, is to be found in Sir Henry Cole's autobiographical account [6].

5.4 Later 'Crystal Palace' designs

The basic techniques of Paxton's masterpiece were applied later to a number of buildings in central London. Outstanding examples are

Figure 5.1   Paxton's design of the Crystal Palace
(HMSO, 1862)

the market halls at Smithfield, Leadenhall and Billingsgate with their cast-iron pillars and high concave roofs. Abroad the design of the 'Crystal Palace' was much copied internationally for later exhibitions, at Cork 1852, Dublin 1853 and 1865, New York 1853, Paris 1867 and Munich 1882. After Henry Bessemer's invention of cheap steel production in 1856, it became possible to replace the cast-iron sections by steel, resulting in an even more elegant structure which was applied in the Paris and Munich designs.

Although the contractual arrangements for the erection of the exhibition building in Hyde Park called for its dismantling in 1852, its use as an exhibition centre was not at an end. The building was re-erected on the eastern slopes of Sydenham Hill in London as a permanent exhibition site. The new Crystal Palace was considerably larger than Paxton's original design. More funding was assured and it became possible to extend the building in ways to suit its new role. The height was increased by two further stories, the central transept

Figure 5.2   Power-operated tools at the site of the Great Exhibition of 1851
(HMSO, 1862)

enlarged, and two end transepts added. Additionally the Palace was flanked by two water towers, designed by Brunel, each 86 m high, to provide a head of water sufficient for the many fountains in the surrounding 81 ha park. All of this increased the weight of the steel structure to 9600 t compared with the 4300 t of the Hyde Park structure.

Shortly after its opening by Queen Victoria in 1854 a railway link was inaugurated to the nearby London to Brighton line, having two serving stations. The connection between these two stations was by

Figure 5.3   The Atmospheric Railway at the Sydenham Crystal Palace in 1854
(Illustrated London News Picture Library)

means of an atmospheric or pneumatic railway, shown in Figure 5.3, and mentioned earlier in Chap. 3. This was a system by which trains were propelled either by air or atmospheric pressure acting on a piston running in a continuous tube. It achieved some success for short runs but when tried out on a large scale was plagued with technical faults and eventually abandoned [7].

The Sydenham Crystal Palace was used for a number of exhibitions, including the prestigious international electrical exhibitions of 1881 and 1891 discussed in Chap. 9, until destroyed by fire in 1936.

5.5 Organisation for the 1851 exhibition

The organising committees for the 1851 exhibition, headed by Prince Albert, included many of the leading scientists of the day. Despite this, engineering and science were not as prominent as they became in later international exhibitions and objects or classes of objects were often selected by their artistic achievement in the Victorian era. Much attention, however, was given to the current public interest in things mechanical, particularly in the newer agricultural machines and applications of the steam engine, both stationary and as a prime mover, so that these inevitably became the major engineering artifacts on display. The Machinery Gallery was extremely popular with the general public. In the previous decade the newspapers and magazines had been full of information on the wonders of the new engineering age and the galaxy of Victorian engineers working on the new inventions. Indeed most of the visitors to the exhibition, even from country districts, were already familiar with the bridges, roads, canals, railways with their huge locomotives, the speed of the telegraph and the new textile and agricultural machines. The railway network was still being developed and trips to London by the new steam trains to see the Great Exhibition' accounted for a large section of the day visitors. Within the gallery the visitors would have been able to see huge marine engines, locomotives, hydraulic presses, agricultural machinery, etc. Steam engines were much in evidence and all told some 60 machines were exhibited [8]. Many of the machines were functional and very noisy, which lent to the excitement. Those not provided with their own motive power were belt driven and coupled to a large mill engine whose boilers were housed in another glass and iron building in the Palace grounds.

Other exhibits in the quieter exhibition areas, also contained mechanical marvels. In a section devoted to civil engineering appeared a model of Liverpool Docks complete with 1600 'fully rigged' ships, models of bridges, a railway layout, a lighthouse and canals. Numerous international committees were formed to choose the exhibits, which eventually numbered over 100 000 from almost 14 000 exhibitors. The exhibition building when finally completed lent itself to a clear division amongst exhibitors with those from Great Britain and her dependencies occupying the Western half and foreign contributions in the Eastern half.

Some 5000 Americans sailed from New York to visit the Crystal Palace and, of course, to see their country's contribution. Although the United States exhibits only comprised 3% of the total displayed in the eastern side (France was allocated the largest area) it did secure more prizes than many continental nations. Prominent amongst these prize-winners were the McCormick reaper, ploughs, firearms—particularly the Colt revolver, adopted later by the British army—and the sewing machine, invented by Elias Howe, an American citizen, in 1846 and not previously demonstrated to a British public [9].

5.6 Classification and selection of exhibits

Before we look in detail at the electrical exhibits, and particularly those in the western side, we need first to consider the way in which the proposed content of the exhibition was viewed. This was influenced very much by the method chosen for cataloguing the exhibits into classes. This caused considerable controversy, expressed publicly by no less a person than Charles Babbage. He had written in 1832 on the classification of industry in his book, '*On the Economy of Machinery and Manufactures*' [10] and was considered knowledgeable in this subject. On this account he was invited by the Chairman of the Classification Committee, Dr Lyon Playfair, Professor of Chemistry at the University of Edinburgh, to assist in the classification discussions, but declined to do so. He was angered by the fact that permission was refused to display his difference engine, although other calculating machines were selected for the exhibition.[2] Of these pride of place was given to Staffel's medal-winning calculation machine shown in the Russian section. This was demonstrated as capable of performing addition, subtraction, multiplication and division and extracting the

---

[2]One of Babbage's calculating machines was, however, exhibited in the second Great Exhibition in 1862.

square root [11]. Other mechanical calculating machines were shown by exhibitors from France, Germany and Switzerland. Babbage was sufficiently upset about his exclusion from the selected list of exhibitors to write about the exhibition in his vitriolic book, *'The Exposition of 1851'* [12], a critique of policies and organisation, not only of the exhibition, but also of what he considered as 'the corrupt state of science in England' at that time.

As a consequence of Babbage's refusal Dr Playfair had to devise his own system. He had considered the system advocated by the French (of which more later) but decided to reject this as *'... an elaborate system of classification based on continental abstraction'*. His mind was made up following a test presented to the French Commissionaire as to who would find it easier to locate a commmon domestic object. He chose a walking stick, and with the Playfair system found this under 'object for personal use' with no difficulty. The French, on the other hand, after extensive searching, located this as 'a machine for propagation of direct motion"! The system devised by Playfair is given below and was to serve, not only for the 1851 exhibition, but for a number of later exhibitions at home and abroad:

1. machines for direct use
2. manufacturing machines and tools
3. mechanical engineering and architectural contrivances
4. naval architecture and civil engineering
5. agricultural and horticultural machines
6. philosophical, musical, horological, acoustics and miscellaneous.

Subdivisions of these classes were made and 30 classes were eventually defined with an international jury assigned in each class. Electrical exhibits were to be found in subclass X—philosophical instruments, with a few exceptions.

Who were the people responsible for choosing and judging the exhibits for subclass X and what were their interests? The jury were selected and were under the guidance of the Royal Commission, of which fully half were members of the Houses of Lords and Commons (including the Rt. Hon. W.E. Gladstone, MP), but it did include three Fellows of the Royal Society and three engineers: John Scott Russell, a naval architect who was to build the Great Eastern, a highly successful cable-laying ship; Robert Stephenson, son of George Stephenson, the railway engineer; and Sir William Cubitt, President of the Institution of Civil Engineers.

The jury for subclass X consisted of: a Chairman, Sir David Brewster,

FRS, chiefly remembered for his work on optics, with James Glaisher, Superintendent of Greenwich Observatory; astronomer Sir John Herschel; W.H. Miller, Professor of mineralogy at Cambridge; horologist E.B. Denison; Richard Potter, Professor of natural philosophy at University College; and an equal number of foreign jurors: Professor Daniel Collandon from Switzerland; Professor Hetsch from Denmark; E.R. Leslie from the United States; Claude Louis Mathieu of the Paris Observatory; Professor Schubarth from Germany; and Baron Armand Seguier from France [13]. We look in vain for engineers and experimentalists familiar with the current state of electrical investigation and manufacture. This is somewhat surprising since very many 'electrical men' were in their productive middle years at the time of the exhibition (see Table 4.1). Mention is made in Bence-Jones, *'Life and letters of Faraday'* [14] that in 1851 Faraday was made *'...a juror of the Great Exhibition'*, but his name does not appear in the lists of jurors for the exhibition and it is possible that he was simply consulted on an exhibition matter. The jury were able to receive advice from Associates, who were not allowed a vote. Since the contents of this subclass on philosophical instruments, which included electrical artifacts, includes apparatus submitted by well known names in the electrical world, such as Joule, Siemens, Henley, Wheatstone, Cooke, Bain, and others, we note that, despite the jury's apparent lack of familiarity in this field, electrical science was found a place in the exhibition, although this was perhaps not as prominent or extensive as one would expect from its progress in the first half of the 19th century.

5.7 The electrical exhibits

5.7.1 The telegraph

In one area, however, the exhibition was particularly rich, and that was in artifacts relating to electrical telegraphy. Whilst the Atlantic had yet to be spanned with submarine cable (this did not occur successfully until 1858), preparations were well under way for this venture. Cables were already in place to link London with Paris and other European capitals, which process was completed with the laying of the Dover–Calais submarine cable in the same year as the Exhibition. This was carried out by Brett Jacob and John Co. who exhibited a model of the printing telegraph later used for this project. These ongoing developments excited a fair amount of public interest in telegraphy apparatus. This was heightened by the spectacular results of telegraph communications, as reported by the penny papers. The Great Western Railway telegraphists, using Wheatstone's five-needle telegraph, were the first to convey news of the birth of the Queen's second son, the Duke of Edinburgh, at Windsor in August 1844, and in 1845 made possible the prompt capture of a murderer, John Tawell, recognised as he boarded a train at Slough station. The news was telegraphed to Paddington, and Tawell was arrested as he left the train.

A problem which had only just been solved was the design of suitable cables, particularly for submarine purposes. The insulation properties of gutta-percha had been described a few years earlier at the Royal Society of Arts in 1843 (see Chap. 4) and several manufacturers at the exhibition displayed examples of this material [15]. Amongst them were the new Gutta-Percha company, formed by Thomas Hancock in 1845 and later to become the Telegraph Construction and Maintenance Co., McNair of Glasgow, Robert Newall, Francis Whishaw (previously secretary to the Society of Arts), C. Walker (who first used gutta-percha as a cable insulation in the line between Dover and Folkestone in 1849), and the Siemens Company of Germany. The insulation properties of gutta-percha were also employed by several exhibitors to construct extremely efficient machines for the production of static electricity. One of these, provided by Westmorland (prize medal awarded) described the rather optimistic possibility of driving his machine by steam power to achieve *'an electric power commensurate with that of lightning'*. In terms of the later achievements of the Van de Graaf Generator, shown at the 1937 Paris exposition, he was not too far wrong.

Telegraph apparatus was to be seen, not only on the stands, but in use about the exhibition. A system had been installed inside the exhibition buildings, its function to warn officials of the presence of pickpockets at large. Elaborate arrangements had been concluded whereby a telegraph message from Colonel Reid to police headquarters would result in the park gates being closed whilst the matter was being dealt with. There is no record of this system actually being used!

Both companies and private experimenters exhibited telegraphy apparatus in 1851 [16]. The Electric Telegraph Company was established in 1846 and held the monopoly of most of Cooke and Wheatstone's patents, including the important one defining the five-needle telegraph, patented in 1837. Many of the exhibits by other companies represented attempts to circumvent this invention and were of lesser utility and commercial success. The telegraphs shown at the exhibition made use of two indicating mechanisms: a Needle

Telegraph using Oersted's observation of the deflection of a magnet-
ised needle under the influence of a current-carrying wire, and an
electromagnet, often employed in conjunction with a ratchet device to
rotate an indicator by means of a series of electrical impulses, or as a
relay to operate a bell or buzzer. Of these the needle telegraphs were
the most numerous seen at the Exhibition; indeed the two-needle
system of Cooke and Wheatstone approached something of a standard
at the time, particularly for railway signalling. The five-needle
telegraph of Cooke and Wheatstone used on the Great Western
Railway is shown in Figure 5.4. Here five wires were required with the

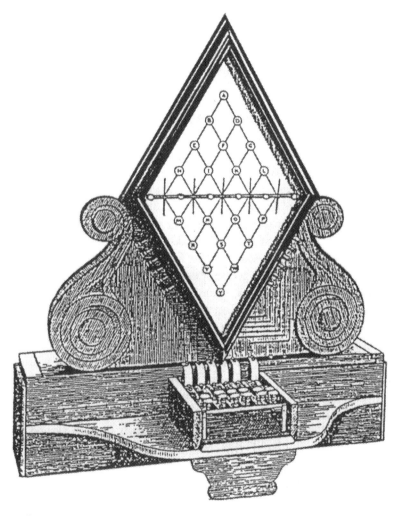

Figure 5.4   The five-needle telegraph of Cooke and Wheatstone
(Official Catalogue of the Great Exhibition, 1851)

current deflecting the needles either to the left or right depending on its direction, the combination indicating the actual letter transmitted. This required some skill to recognise the needle combinations and needed a trained operator. This was limited to transmitting only 20 letters of the alphabet and presented a problem to the telegraph operator, mentioned earlier, responsible for transmitting Tawell's description as, '*a man in the garb of a Quaker*'. With commendable resourcefulness he promptly sent the word KWAKER, which was duly recognised at the receiving end.

One concern for needle telegraph operators at the time was the possibility of atmospheric electricity or lightning discharge affecting the instrument. Highton exhibited his patented 'lightning strainer' to deal with this contingency. The circuit wires, covered only with bitumous paper, are caused to pass through a box of iron filings, connected to the ground. Whilst the insulation is adequate for the weak telegraph currents, a lightning discharge would be expected to puncture the paper and bypass the discharge to ground.

Other complex systems using needles were shown by Gauss, Weber, Mason and Galton. Some made use of a simpler indication, with one needle for each letter, but these required a large number of lines for their operation—30 in the version exhibited by William Alexander, and a similar number by Schilling of Austria. The Alexander display was claimed to be the first ever exhibited, being shown earlier at the Royal Scottish Society of Arts in 1837 (see Figure 4.3). It is interesting to note that Alexander used a keyboard, '*precisely similar to that of a pianoforte*'—an adaptation which was followed by a number of later investigators [17]. These early telegraphs, sometimes referred to as 'speaking instruments', were so called because the needle(s) indicated the letters of the alphabet spelled out laboriously one at a time. One of these, the Siemens Pointer Telegraph, was awarded a Council Medal at the exhibition. The Morse instruments adopted later enabled a continuous stream of characters to be transmitted and interpreted by the operator.

Two alternative techniques requiring only a single line were much in evidence. Samuel Morse had publicised his dot-dash alphabet in 1844 which was capable of being related to the two possible directions of current along a line and hence to alternative deflections of the magnetic needle. The principle was also used to obtain an audible signal from a buzzer. Both needle and buzzer indication required the employment of a skilled operator. It was probably one of these coded transmissions that was demonstrated to Queen Victoria during one of her several visits to the Exhibition. She commented in her diary,

'... *July 9th; We went to the Exhibition and had the electric telegraph show explained and demonstrated before us. It is the most wonderful thing and the boy who works it does so with the greatest of ease and rapidity. Messages were sent out to Manchester, Edinburgh etc. and answers received in a few seconds—truly marvellous! Colonel Wylde was there to explain all to us'* [18].

Not all the exhibitors using a single pair of lines made use of a Morse code sender. Much in evidence were the devices using a relay to control a ratchet to actuate a circular plate engraved with the letters of the alphabet. The sender first set the dial to the letter required through a crank handle. This actuated a clockwork mechanism which, through a ratchet, sent out a series of electric pulses via a switch and battery, corresponding in number to the set position of the dial as the clockwork mechanism ran down. The pulses passed along the transmission line to operate a relay actuated ratchet in the receiving device which was made to rotate an indicating dial accordingly, returning to its initial state again by a clockwork mechanism. Several devices of this kind were shown by Wheatstone, Barlow, Nott & Gamble and Bréguet. The French Bréguet device, shown in Figure 5.5, was more complex, requiring two cranks to simulate the indications required by the Chappe system, already in use throughout France. This was a mechanical semaphore signalling system, invented by Claude Chappe in 1793 and claimed to be the first visual signalling system. It made use of two possible positions of six signalling discs and hence could provide 64 symbol combinations, accommodating not only all the letters of the alphabet but a number of other characters as

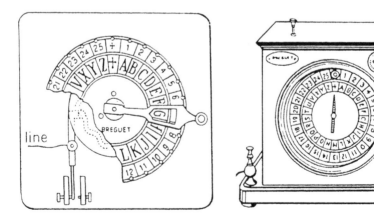

Figure 5.5   Bréguet's Alphabetical Telegraph
(Official Catalogue of the Great Exhibition, 1851)

well. It was introduced initially in 1794 to connect Paris to Lille during the Napoleonic wars. The Chappe code was applied later in all the French electrical telegraph systems exhibited in 1851, '... *in order to avoid retraining of their semaphore operators*', and in the process retarding the development of telegraphy in France for several years [19].

The ability to transmit a series of pulses over a single pair of lines encouraged a number of inventors to show their 'printing telegraphs', which use the relay effect to bring a pen down on a moving paper strip, thus making the transmitted Morse code visible. A printing telegraph, shown by Siemens & Halske, was awarded a Council Medal. A system of this kind was later to be adopted for submarine cables and, following the successful spanning of the Atlantic, a number of printing telegraphs were shown at the 1862 Exhibition (see later). Alexander Bain exhibited a chemical printing telegraph whereby the signals are caused to impress and mark on a paper strip, impregnated with prussia-muriate, a bright blue line when drawn past the actuating relay by means of a clockwork mechanism. (An earlier version was exhibited at the Polytechnic Gallery in 1842.) Bain's chemical system was also used in modified form in the transmission of previously prepared diagrams and this invention is generally considered to constitute the first working facsimile equipment [9]. Bain patented his system and eventually sold it to the Electric Telegraph Company for £7000. It was, however, not the only prototype facsimile equipment exhibited. Fred Bakewell, 'inventor and patentee', demonstrated a remarkable copying electric telegraph for the transmission of handwritten signatures, shown in Figure 5.6, which he visualised as being applicable to banking operations, anticipating credit-card transfers by more than a century.

5.7.2 Motors and generators

To the engineers of the Victorian era a 'prime mover' meant a steam engine, and the notion of an electric motor was not a respectable one. However, the idea was abroad and actually achieved by Thomas Davenport in 1837 in the United States, who developed a motor weighing 23 kg, running at 450 rev/min and capable of drilling holes in sheet steel. Davidson, a Scottish engineer, had incorporated a motor of his own design in an electric locomotive and shown this to the Royal Scottish Society of Arts in 1842 (see Chap. 4), and finally a few years earlier in 1838, Professor Jacobi of St Petersburg had actually propelled an electric-powered boat carrying 14 passengers at almost 5 km/h on the river Neva [20]. However, none of these delights faced

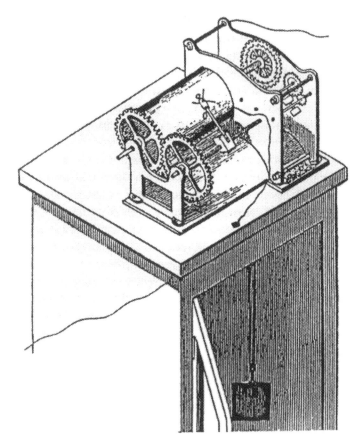

Figure 5.6   Bakewell's Copying Electric Telegraph
(Official Catalogue of the Great Exhibition, 1851)

visitors to the Great Exhibition. This may have been due to the outspoken views of I.K. Brunel (one of the influential jurors), who deliberately told Scott Russell, when he was making the first plans for the exhibition, that he did not wish to hear '... *any mention of electric machines which can as yet only be considered as toys*' [21]. As late as 1859 well known scientific figures were to hold firmly to this view. Professor Rankine in the preface to his book on *'The steam engine'* writes,

'... *their small importance, as prime movers and absence of economy ... the true practical use of electro-magnetism being not to drive machinery, but to make signals (through telegraphy)'*.

By 1851 the general principles of reciprocating electrical engines, such as W.C. Henry's (which was shown at the Exhibition) and the more successful rotating motors of William Ritchie, who had earlier demonstrated his machine at the Royal Institution, were well known. The majority of experimenters at the time were, however, working on devices having a reciprocating motion, using a crank to obtain rotary action. One of the more successful of these was that due to Soren Hjorth from Denmark (a prize-winner). This relied on simple in-line magnetic attraction to convert oscillatory motion into continuous rotary motion by means of a crank and piston. A soft iron piston, attached to a crank, was attracted by the action of a coil energised through a battery. At the end of its stroke the coil current is reversed and the piston action continues in the reverse direction [22]. Hjorth went on to develop electrical generators which he patented in England, and he very nearly succeeded in inventing the self-excited generator, a major step in power generation which was to await Wheatstone's and Siemen's work in 1867 (see Chap. 8) [23]. Other similar reciprocating motors were demonstrated by Harrison, Knight and Watkins, but none produced sufficient power to drive any useful device.

The only electric prime movers doing what Brunel would have called 'useful work' were the various electric clocks shown. Alexander Bain, who was responsible for the extremely successful printing telegraph noted earlier, exhibited his design, and there were several others. However, pride of place went to Charles Shepherd of Leadenhall Street, who provided the highly visible electric clock in the Great Transept to the Exhibition. The dials were of unusual semicircular shape and 1.5 m in diameter (Figure 5.7). Two further clocks were installed, one in the transept and another at the western entrance, and were both wired to the parent mechanism in order to keep in synchronism with one another.

The power for these clocks was provided by a battery of Smee cells chosen for their simplicity and ease with which they may be recharged. ('Recharge' at this time meant the replacement of the zinc plates within the battery—the electrically rechargeable cell, and, indeed, a

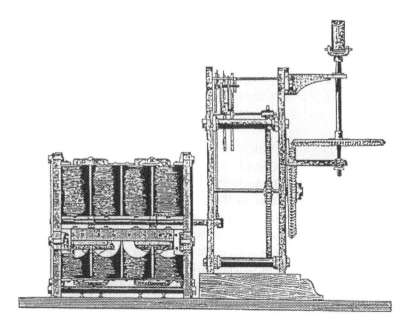

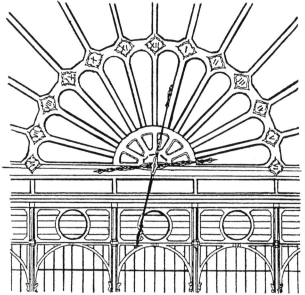

Figure 5.7   *Charles Shepherd's Electric Clock*
(Official Catalogue of the Great Exhibition, 1851)

convenient way of carrying this out, had yet to be developed.)
Batteries, or 'galvanic batteries' as they were then called, were the
main source of continuous electrical power at that time. A variety of

cells developed for different purposes were shown at the Exhibition: the Smee zinc/silver cell; Grove's zinc/platinum cell; Daniell's standard zinc/copper cell; Cooke's 'telegraph' cell, also using zinc/copper plates; Walker's graphite battery; and several others (the ubiquitous rechargeable lead/acid cell was not invented by Planté until 1860). A major manufacturer of batteries at the time, George Knight of Foster Lane, London, exhibited all of these types. There was a singular amount of mysticism about at this time concerning electrical power derived from batteries. One worthy doctor, G. Meinig of London, exhibited his portable galvanic battery chain, described as

> *'... consisting of the metallic combination of various galvanic elements; so arranged as to be very portable ... Designed to be worn on the body for the purpose of effecting the cure of various chronic diseases by means of the electric current, which in its passage from one pole to the other, passes through that part of the body encircled by the chain'* [24].

Electromagnetic sources of electrical power had been known since the work of Pixii, Saxton and Clarke in the 1830s. Similar rotary electromagnetic engines were shown by Stanton, Hardy, Dunn and Simons but contributed nothing new to the earlier designs. Small generators of this kind were often used for telegraph work. These were hand-operated, to produce pulses of current to energise a galvanometer indicator at the far end. William Henley exhibited one of these as part of his patent telegraph apparatus. In another version, operated by a treadle, continuous current was made available, using a commutator.

Closely allied to these continuous generators were the various 'galvanic apparatus' or 'shocking coils' on display. One ambitious version of this genre was equipped with

> *'... a graduated regulator employed to administer galvanic electricity ... which regulated intensity of shock by two moveable indices of power ... it is the invention of Mr Hearder of Plymouth and is intended for the use of private families'.*

5.7.3 Other electrical exhibits

An entire subclass 23a was devoted to examples of electroplating, the first of the industrial applications of electricity. Moritz Jacobi invented electroplating in 1837 and subsequently published a book about the technique. However, it was the Swiss who successfully sought for and obtained a patent for carrying this out on an industrial scale, and only licensed the technique to John Wright of Birmingham in 1840, who initiated the plating industry in Great Britain. Originally the power source was from batteries which proved expensive, but by 1842 John Woolrich had obtained a patent relating to electroplating using a magnetoelectric generator, and this was used by Elkingtons, another Birmingham firm, to carry out electroplating on a commercial basis [22]. It was the firm of Elkingtons which produced the most elaborate (and largest) collection of electroplated articles at the exhibition, including a jewel-case and several elaborately decorated tables, the property of Queen Victoria. No examples of the actual process of electroplating were shown. With the availability of continuous electrical power from a reasonably efficient generator, which occurred after 1870, the process became widespread and a new range of industrial applications, such as the production of galvanised iron sheeting, became possible.

Several electromagnets were shown. The most significant was that of James Prescott Joule. It weighed over 50 kg and used a battery 'of moderate power' to enable it to support a weight of >1 t. Rühmkorff showed some magnetic instruments, including early samples of his efficient induction coils or transformer. Measuring instruments displayed included galvanometers, a magnetometer, anemometer and electrodynamometer, and an electric measuring instrument for physical length measurement submitted by Joseph Whitworth, the mechanical engineer from Manchester.

A number of arc lamps were shown at the Exhibition, incorporating early forms of self-regulating systems to progressively advance the carbon as it became consumed and so maintain the arc. William Staite and Petrie, pioneers of electric lighting in the Victorian era, had been experimenting with control equipment for carbon arcs for some time and by 1851 had filed a number of patents for their designs [25]. In May 1851 Staite wrote a long letter to *The Times* in which he claimed that the arc-lamp regulating devices shown at the Exhibition were based on his invention, but the letter was never published [22].

Little attention was paid to electric lighting at the Exhibition. It was generally considered impractical to realise this as a possible application of the new electrical power. George Virtue, writing in *The Art Journal* of 1851, categorically asserts a common view that

> *'Although satisfied with our present knowledge of electrical forces, we can scarcely hope to adapt the electric light [arc light] to any useful purpose, within the limits of any ordinary economy'.*

There were in fact, many attempts to produce a satisfactory regulated carbon arc lamp between 1844 and 1859, of which those of Staite and Petrie were perhaps the most successful, but all were defeated by the limitations in battery power and the heavy consumption of carbon rod during their operation [26]. In the case of the Great Exhibition there was another reason for the lack of interest in electric lighting installation, and this lay with the exhibition building itself. Its all-glass construction was in part designed to make use of natural lighting during exhibition hours (it was not open in the evening). Artificial light was not considered necessary, although a limited amount of gas lighting was installed.

5.8 Electrical exhibits from abroad

The total number of foreign exhibitors, those located in the eastern half of the building for foreign exhibitors was roughly equal to the number of British exhibits in the western half. The number of electrical exhibits was, however, smaller. Whilst the exhibitors of electrical artifacts from Britain and its dependencies reached 41 out of a total of 7000 exhibitors, France provided only 9 out of a total of 1740, Germany, 11 out of 1524, and America, six out of 566. All these figures completely underrepresent the considerable effort being carried out at the time in all industrial countries to apply the recently discovered principles of electricity to industrial and domestic devices [27] and relates more to the activities of the selection committees than to the state of electrical knowledge and activity of the time.

Several of the foreign contributions have already been described. A few were unique to the exhibition as representing a new technique or invention. One of these was a very large thermoelectric battery, applying Seebeck's principle to the production of electricity by the application of heat. This was shown by Süss of Prussia and consisted of a central ball of heated iron some 60 mm in diameter radiating on to five thermoelectric couples, each contained in a hollow brass cylinder. The thermoelectric elements were electrically linked together to provide a complete circuit with the current produced used to power an electromagnet or to provide a chemical reaction (presumably electrolysis). Another was an 'electromagnetic self-registering anemometer' of Froment (France). Here the four cups of the anemometer each carry a magnet revolving within a coil of wire. The current produced is related to the wind velocity, and taken together the four currents produced could also indicate the direction of the wind.

5.9 Aftermath of the Great Exhibition

Despite the elaborate arrangements for choosing the exhibits with the many subcommittees overseen by the great and the good of Victorian society, some remarkably useless devices were exhibited to the public. These included a silent alarum bed which tipped the sleeper unceremoniously on to the floor at a prearranged time, a false nose made of silver and a penknife with 80 blades. Nor were the electrical exhibits free of this folly. A galvanised walking stick was shown which gave a slight electric shock if held in one hand and a violent one if held in both hands, and also a 'comic telegraph', shown in Figure 5.8, which took the form of a man's head, where symbols corresponding to keys remotely pressed appeared on flags seen at the top of the head, whilst the mouth moved meaninglessly. A long-standing critic of the Exhibition, Colonel Sibthorp, who had opposed the creation of the Palace in Hyde Park from the beginning, was moved to remark on, '... *trumpery and trash in abundance*', and certainly amongst the minor exhibits were a number, selected for the most bizarre of reasons. With hindsight we can note that the Victorian admiration for painstaking work, often using the most intractable materials, and love of elaborate decoration for its own sake, did result in galleries filled with objects of little practical use, whose only function seemed to be to evoke wonder from the visiting population [28].

Viewed as a whole the Exhibition was, nevertheless, a great success and served to place on record the achievements of a generation of skilled craftsmen, engineers and designers. Henry Cole was to write, after the Exhibition,

> 'The history of the world, I venture to say, records no event comparable in its promotion of human industry, with that of the Great Exhibition of the Works of Industry of all Nations. A great people invited all civilized nations to a festival, to bring into comparison the works of human skill'.

It was indeed truly great, not only for its location, number of exhibits and appeal to visitors but also as a financial venture. Its profits enabled the purchase of a large tract of land in Kensington, some 20 ha in extent, and the construction of several museums and other buildings on the site, including the Victoria and Albert Museum, the Science Museum, and the Imperial College of Science and Technology, which has had a profound influence on science, engineering, electrical research and teaching for more than a century. Apart from its

Figure 5.8   The 'comic' telegraph
(Official Catalogue of the Great Exhibition, 1851)

coverage of telegraphy, however, the Exhibition did not describe the then current state of electrical progress in the 19th century at all well. Out of a total of about 14 000 exhibits <1% were even remotely related to electrical engineering, and many of these were displayed in the least conspicuous areas of the exhibition. However, the rapid development in invention and application in the second half of the 19th century more than made up for this, so that by the post 1870s huge demonstrations of electrical prowess through international exhibitions were made—as we shall see later.

5.10 References

1 HUDSON, D., and LUCKHURST, K.W.: 'The Royal Society of Arts, 1754–1954' (John Murray, London, 1954)
2 TALLIS, J.: 'Tallis's history and description of the Crystal Palace', 3 Vols. (Tallis, London, 1851)
3 HOBHOUSE, C.: '1851 and the Crystal Palace' (John Murray, London, 1950)
4 KIHLSTEDT FOLKE: 'The Crystal Palace', *Sci. Am.*, Oct. 1984, pp. 132–43
5 WYATT, M. DIGBY: 'Illustrated catalogue for the Great Exhibition of 1851' (Royal Commission, London, 1851)
6 COLE, SIR HENRY: 'Fifty years of public works', 2 Vols. (G. Bell and Sons, London, 1884)
7 HADFIELD, C. 'Atmospheric railways' (David & Charles Publishers, Newton Abbot, 1967)
8 FFRENCH, Y. 'The Great Exhibition 1851' (Harvill Press, London, 1950), p. 213
9 BURNS, R. 'Alexander Bain, a most ingenious and meritorious inventor', *IEE Eng. Sci. Educ. J.*, 1993, **2** (2), pp. 85–93
10 BABBAGE, C.: 'On the economy of machinery and manufactures' (Charles Knight, London, 1832)
11 'Staffels calculating machine', *Illustrated London News*, July–Dec 1851, **19**, p. 356
12 BABBAGE, C.: 'The Exposition of 1851' (John Murray, London, 1851)
13 BENNET, J.A.: 'Science at the Great Exhibition' *in* 'Exhibition catalogue for the Whipple Museum' (Whipple Museum, Cambridge, 1983), p. 2
14 BENCE JONES: 'Life and letters of Faraday, Vol. II' (Longmans Green, London, 1870), p. 287
15 EUGENE OBACH: 'Cantor lectures', *J. RSA*, 1856, **XLVI**, p. 98
16 'Official catalogue of the 1851 Exhibition', 3 Vols. (Spicer, London, 1851)
17 DAWSON, K.: 'Electromagnetic telegraphy. Early ideas, proposals and apparatus', *in* HALL, A., and SMITH, N. (Eds.), 'History of technology, Vol. 1' (Mansell, London, 1976), pp. 113–141
18 FAY, C.R.: 'Palace of industry' (Cambridge University Press, 1951), p. 63
19 DAUMAS, M.: 'A history of technology and invention, Vol. III', 1725–1860 (John Murray, 1980), p. 383
20 'Jacobi's boat', *Philos. Trans. R. Soc.*, 1839, **15**, p. 164
21 NOBLE, C.: 'The Brunels, father and son' (Cobden-Sanderson, London, 1938)
22 'Historic machines—Soren Hjorth's first machine', *The Electrician*, 1882, **9**, pp. 173–175
23 BOWERS, B.: 'A history of electric light and power' (Peter Peregrinus, London, 1982), pp. 83–84
24 'Official catalogue of the 1851 exhibition report of the Juries' (Wm. Mackenzie, London, 1851), pp. 621–629
25 WOODWARD, G.: 'Staite and Petrie: pioneers of electric lighting', *IEE Proc. A.*, 1989, **136** (6), pp. 290–296
26 BRIGHT, A.A.: 'The electric lamp industry' (Macmillan, New York, 1949)
27 MACLAREN, M.: 'The rise of the electrical industry during the nineteenth century' (Princeton University Press, 1943), p. 94
28 ROLT, L.T.C.: 'Victorian engineering' (Penguin Books, London, 1970), p. 159

Chapter 6
British exhibitions after 1851

6.1 National and international exhibitions

1851 was a watershed in the history of industrial exhibitions; before this date all exhibitions were national; after it they were almost all international, even when held away from the capital of the country concerned. They also became larger and more flamboyant until the showmanship threatened to overwhelm the information content of the event—but an examination of this lies many chapters further on.

The success of the Great Exhibition was not, however, to be repeated in Britain during the second half of the 19th century. Indeed it is regarded by many as representing the peak of achievement for the industrial revolution in Britain and that *'... after 1851 things would never be quite the same again ...'* [1].

A number of significant exhibitions on different themes were held in Britain during the remainder of the century, including an extremely successful series of industrial exhibitions to celebrate the Jubilee of Queen Victoria in London and provincial towns. However, for the large comprehensive exhibition showing the range of industrial power becoming available in the international community we need to look to France and the United States, and this will be the subject of the next Chapter. Here we will first examine the technical content of the 1862 exhibition held in Kensington, which was in many ways much more rewarding than the 1851 exhibition had been in terms of the technology presented, but failed in its wider aim to attract international support.

6.2 The 1862 International Exhibition

In 1862 the second Great International Exhibition was held in a new venue in Kensington (on the site of the present-day Natural History Museum). The building was an immense brick erection consisting of two vast domes of glass, 76 m high and 18 m in diameter, connected by a nave some 244 m long and 30 m high with its upper portion lit through a range of clerestory windows. The area covered was ~6.5 ha with two annexes providing an additional 3 ha.

It attempted to repeat the earlier success of the 1851 Exhibition, but although the attendance was very similar it did not command such widespread interest. Major innovations displayed were the new Bessemer process for producing steel, Babbage's calculating machine, part of which was finally exhibited, having been refused for the 1851 Exhibition, and significantly, Ferdinand Carré's ice-making machine, the forerunner of the domestic refrigerator. This consisted essentially of a boiler, partly filled with ammonia, standing on a portable stove and connected to a small conic vessel, having double walls. The process of liquefaction of the ammonia gas caused the water contained in the smaller vessel to freeze; 1 kg of ice was produced in 2 h [2]. The Bessemer process was of great interest to industrial visitors, particularly those representing German heavy industry, and its demonstration in 1862 is often credited with the rapid spread of steel-making in Germany which followed.[1] By the time of the 1867 Paris exhibition, the Krupps Company of Prussia was displaying a 50 t steel cannon capable of firing 454 kg shells, a major armaments advance for its time, and the makers were awarded a grand prize for the innovative methods used in steel production which made such designs practicable.

The exhibited machinery in 1862 was, however, even greater than in 1851. Some 1100 machines were shown (not all of them working). These included several locomotives, cotton machinery, brick-making machinery, static steam engines, road carriages, marine engines, workshop machines and a small exhibit of magnetoelectric machines [3]. Some of the industrial contributions, first shown at the 1851 exhibition, were much improved and indicated the continued

---

[1]The Bessemer process, discovered in 1856, required nonphosphoric ores which were freely available in England but less so in the mainland of Europe. By 1879 Gilchrist Thomas had demonstrated a process whereby phosphoric ores could be used, and it was this latter process which allowed Germany to dominate the world trade in steel by the 1900s

strength of British industry at the time. One of these was an impressive demonstration of Platt's mule, a combination of Arkwright's spinning machine and Hargreaves' jenny, which completed the process of yarn production from plucked cotton. A display of 'machinery in motion' was shown with immense working steam engines, hydraulic presses, ships' engines and locomotives—some in the grounds and others enclosed by a formidable structure of iron girders and roofing trusses. William Fairbairn, the Manchester engineer, noted the considerable progress that had been made since 1851 in the machine tools exhibited which now enabled tolerances of 0.025 mm (0.001 in) to be realised.

There was evidence at the 1862 Exhibition, however, of an important change in direction which British exhibitions were taking at this time, and which had an effect of reducing the impact of this display of manufacturing prowess. In competition with the French, much emphasis was placed on the arts, and a large picture gallery now formed part of the collection exhibited. The fine arts displayed were enthusiastically presented and followed a theme of *'progress and present condition of the arts in England'* which extolled the English national school of art at the expense of foreign exhibits and thereby received much press support.

6.3 The electrical exhibits

For the smaller technical displays manufacturers were requested only to submit new equipment, *'constructed since the 1851 exhibition'*, which appears to have limited the range of electrical artifacts shown. As before, telegraph apparatus formed a large part of the exhibition of electrical artifacts. By 1862 there were over 24 000 km of telegraph cables laid in Britain, endowing even greater significance to the technology. An ephemeral success had attended the spanning of the Atlantic by telegraph cable in 1858 and efforts were under way to continue with the project, together with a grand design to link London with Calcutta with a (mainly) overland telegraph route [4] so that the marvels of long-distance Morse telegraphy were. becoming well known to the exhibition-going public. (We may also note that for the 1862 Exhibition Charles Wheatstone had been appointed to the organising committee for this section and also acted as one of the jurors.)

A visitor to the exhibition was confronted by the practical results of telegraphy at the entrance. John Timbs of the RSA writes [5],

> '... at the very threshhold of the Exhibition.... the 'magnetic teletale'
> of Professor Wheatstone was attached to some of the turnstiles, and
> this, in a measure, controlled the financial department. This
> instrument was worked without battery power of any kind. The
> electricity was generated by a peculiarly constructed magnetic-
> machine, so connected with the axis of the turnstile as to discharge
> a current of its force at each revolution of the stile. Thus, each
> visitor, on passing through it, unconsciously and telegraphically
> announced his or her arrival to the financial officers in whose rooms
> were fixed the instruments for receiving and recording the liberated
> current, which latter was conducted thither by a line of copper wire
> ....The registers , thus obtained formed a complete check upon the
> money taken at the doors...'

Was this the first example of effective remote data processing ?

A number of manufacturers showed specimens of cable used in the
English Channel, Atlantic and Red Sea routes, including the Malta to
Alexandria cable which had only just been laid in 1861 by Glass, Elliott
and Company, using the new gutta-percha insulation. Other
companies were W.T. Henley, T. Allen, R.S. Newall and the German
firms of Felten & Guilleaume, and Siemens & Halske [6]. Siemens
showed a telegraph wire-covering machine used for submarine cable
production. A London firm of A. Smith Ltd also exhibited machinery
for cable-making. Needle telegraphs were confined to those of
Wheatstone and modifications shown by competing companies. Many
variations of alphabetical and step-by-step telegraphs were exhibited
by competing countries in which the clockwork mechanism seen in
1851 was dispensed with and, in many cases, the battery and switches
also.

An important system shown was that of Siemens & Halske, already
in use in much of northern Europe, including Russia. An extensive
telegraph system was in operation in Russia, extending from the Black
Sea port of Odessa to Moscow, St Petersburg and Finland, with a
further train of stations to Poland. The Siemens pointer telegraph
consisted of a brass dial with engraved alphabetical letters on its
circumference. It was set to the desired letter to be transmitted, and a
handle rotated actuating an electromagnetic generator which then
produced a series of pulses corresponding to the dial position,
transmitting these to effect a similar dial indicating device at the
receiving end. It was slow but effective and required a minimum of
operator training when compared with earlier systems.

Other alphabetical telegraphs were shown by Wheatstone, Bréguet
and others. The printing telegraph of Jacob Brett was exhibited by

which the first message was received through the first submarine cable. Charles Bright's acoustic telegraph, extensively employed by the British and Irish Magnetic Telegraph Company, was shown. This consisted of two electric bells of different tones which respond to negative or positive pulses, respectively, and hence enabled an operator to distinguish aurally the dots and dashes of the Morse code (although the duration of the bell notes are the same). This was claimed to be very rapid in action but was subsequently overtaken by various paper recording machines. A recording telegraph, using Morse code and perforated paper, had been developed commercially with the expansion of the telegraph network and was shown by British, French, Swiss, Austrian and German companies. One British system, designed by Alexander Bain, was able to increase the transmitting speed from ~50 words/min, achieved by a trained operator, to some 400 words/min. Ink recorders were also shown; that of Morse, exhibited in the American section of the exhibition, is seen in Fig. 6.1. Telegraph lines had been laid in 1849 between Vienna and Berlin and the German–Austrian Telegraph Union formed. Morse's system had earlier been demonstrated in Vienna in 1845. Siemens & Halske exhibited a large number of direct writing ink recorders which were at that time widely used in Germany and throughout Europe.

To demonstrate the operation of a telegraph station the Electric and International Telegraph Company arranged to have a number of stations situated within the exhibition building. From these messages could be sent by visitors to, *'all the telegraph stations of the world'* [4]. A charge was made for these and, not to be outdone, a rival company, not able to obtain a foothold within the exhibition grounds, set up its own station on a piece of vacant ground in the Cromwell Road and proceeded to provide a telegraphic service at a lower rate, using their own wires and, via connection to those of the Magnetic Telegraph Company, to *'all British and Continental stations'.*

The remainder of the electrical exhibits were less imposing. Electro-magnetic engines were again derided in terms expressed by Robert Mallet, CEng, FRS, FGS, as,

> '... *we may almost set aside here electro-magnetic engines as a source of power, for we are not aware that any important examples of such have been exhibited. Those electro-magnetic machines shown in the Western Annex and the British and Foreign department are not designed as sources of power but of light, which is here evolved along with heat, by the expenditure of mechanical motion in producing reactions between magnetism and current electricity'* [6].

Figure 6.1   Morse's ink recorder
(Mollet, 'The International Exhibition of 1862')

Despite this negative view a number of electromagnetic engines *were* displayed and operated as power sources for the new electric lighthouses coming into operation. One shown by Frederick Holmes (Figure 6.2) included a commutator, and was driven by a steam engine. It was described as being ' ... *similar to the apparatus used in the Dungeness Lighthouse'* but was probably much smaller since the Dungeness installation was the biggest Holmes had designed, with the generator having a diameter of 2.4 m and weighing more than 3 t. This had only been installed in the year of the exhibition.

Holmes had earlier placed proposals for substituting generators for chemical batteries in lighthouses before the Corporation of Trinity House in 1857. Michael Faraday, as scientific advisor to the Corporation, was consulted with the result that the Holmes proposals were first tested across the Thames at Woolwich and pronounced to be very successful. With Faraday's support the system was given a full-scale

Figure 6.2   Holme's lighthouse generator
(Science Museum photograph)

trial at the South Foreland Lighthouse. The generator used was massive for the job it had to perform, weighing 2 t, and was extremely inefficient. When driven by a steam engine at 600 rev/min the power output obtained was <1.5 kW. However, it proved very practical for a difficult fixed installation, such as a lighthouse, and the idea was taken up in several permanent installations in England and in France. In France the generators were De Meritens alternators, designed by the Compagnie de l'Alliance and installed initially at the La Héve lighthouse in 1863 [7].

Two further generator machines were shown—one from France and another from Siemens—both providing power for electric arc-lamp apparatus. The Siemens exhibit was noteworthy in that it incorporated his shuttle-wound armature, developed in 1856, initially for AC generation, and exhibited here in one of his later DC generators. A reviewer comments on the low efficiency of these generators (10%), in conversion of coal to steam, to mechanical rotation, to electricity and finally to light.

Lighting now formed an important feature at the 1862 Exhibition showing considerable advances since 1851. A major reason for this is that after 1857 various arrangements of self-adjusting arc lamps began to become available commercially and a number of arc lamp displays were shown. An interesting new light display was given by Prosser of a 'lime-light' apparatus, later known as the Drummond light. This functions by heating a piece of lime to incandescence in an oxy-hydrogen flame. Professor Floris Nollett had earlier proposed a system whereby these gases could be obtained by electrolysis and so indirectly drive the lamp by electricity [8].

There was still considerable interest shown in the production of static electricity for experimental purposes, and a number of electrostatic machines were shown. Elliott exhibited a large 'electrifying machine' with ebonite discs. Varley showed the largest machine at the exhibition, with ebonite discs 1 m in diameter and producing sparks of 40 to 50 cm in length. Other manufacturers used the properties of gutta-percha for this purpose.

The widespread and rapid development of the telegraph network required the use of line and insulation testing instruments, and several of these were exhibited by Wheatstone, Varley, Siemens & Halske and others. This section included galvanometers by White and Siemens, the Varley bridge, and a display of the instruments used at Kew Observatory, which included Thompson's electrometer for measuring atmospheric electricity. Also shown was an interesting exhibit by Bonelli of an electrically operated silk loom. This employed Jacquard-holed plates allowing contact through the holes to operate relays controlling the weaving frame.

6.4 Later British exhibitions, 1870–1890

Although Britain won many of the prizes at the 1862 International Exhibition, thoughtful observers were alarmed by the evidence of competition from abroad; it was clear that Britain's easy supremacy in trade and commerce was threatened. Many considered that other countries were overtaking her because their craftsmen were better educated and more liberal support was given to technical and scientific research [9]. That this was also the situation in British universities was noted by Justus von Liebig, the German chemist, who stated,

> '...*That it is a requirement of our times to incorporate the natural sciences as a means of education into the University courses is not perhaps doubted anywhere except in England*' [10].

It was perhaps also unfortunate for Britain in the period between 1851 and 1862 that the French exhibitions should be contrasted with the British in terms of Anglo–French rivalry [11]. This view was publicly stated by Beavington–Atkinson in April 1862, namely

> '...*In these International Exhibitions the contest has been, and will be, mainly between those noble foes, now happily firm friends and allies—England and France ... take likewise front rank in the world's arena of arts and manufactures*' [10],

and also that of M. de Neufchâteau, French Minister of the Interior, at the opening of the French 1862 Exposition, who regarded this as '...*an episode in the struggle against British industry*'.

Whether it was the effect of these public views, or the general gloom-laden atmosphere of the Britain of 1862 (the death of Prince Albert from typhoid in 1861, the Lancashire Cotton Famine and the effects of the American Civil War), the official enthusiasm for international exhibitions waned. Another 'Great' international exhibition for 1871 was proposed but was not carried out. Instead Henry Cole put before the Commissioners a scheme for holding a series of annual exhibitions of 'Selected works of fine and industrial arts' to cover the decade from 1871 to 1880, each one dealing with a specific industry [12]. It is possible, of course, that at the time industrialists and parliament did not feel the need for any more elaborate undertaking since in 1870 the foreign trade of the United Kingdom and her colonies exceeded that of Germany, France, Italy and the United States put together [13]. If so it proved a short-term reaction.

The first of these 'selected works' exhibitions in 1871, held in the same South Kensington venue as the 1862 exhibition, contained four classes: fine arts; scientific inventions and discoveries; manufacture; and horticulture. This was fairly successful, with 33 foreign exhibitors taking part and over 7000 industrial exhibits shown. The exhibitions for the rest of the decade were less effective and only the first four were held. They included a varied selection of industry ranging from civil engineering to cookery. All four contained a separate class on inventions and discoveries but were not notable for their electrical exhibits. It is interesting to note, however, that had the proposal for the year 1876 been carried through this would have included a separate section on electricity and perhaps would have anticipated the success of the technical content of the Paris International Exhibition of 1878.

The 1872 exhibition contained a section on music which included an interesting electrical exhibit and demonstration by Hermann von Helmholtz. In his seminal work, 'Sensations of tone', published in 1863 [14], he showed that vowel sounds could be built up by the combined effects of a group of tuning forks, each vibrating at its own frequency. At the Exhibition he showed how electrical impulses could be obtained from each tuning fork by attaching a contact breaker, connected to a battery, to one of the arms of the fork. The resulting pulses could then be transmitted to an electromagnet attached to a second tuning fork, at the receiving end, to give out the same musical note. Excitation of several tuning forks in this way resulted in the recreation of the transmitted vowel sound. This demonstration was part of early investigations leading to the invention of the telephone by Graham Bell in 1875, but it also led directly to the 'harmonic telegraph'. Bell proposed the use of a number of tuning forks, each producing a series of electrical impulses at its own characteristic frequency, which would be combined and simultaneously sent down a telegraph line to activate a similar set of forks at the receiving end. Associated with each transmitting fork would be a telegraph operator sending a separate message down the line which would be recognised only by the appropriate fork at the receiving end. The achievement of a form of frequency-division multiplexing and transmission in this way echoes one of the principle methods of information transmission in use today, on a worldwide basis, by all telecommunication authorities.

The 1873 Exhibition was concerned with fine arts and a miscellaneous group of industries manufacturing silk, steel, surgical instruments, carriages, food and cooking apparatus. It also had a section on inventions. These four exhibitions were popular, with about 500 000 visitors attending the 1873 event.

After the 1873 Exhibition the scheme for annual exhibitions in Great Britain was dropped, to be revived in 1883 when a further proposal was made to hold separate exhibitions on specific topics, namely

1883 fisheries
1884 health and education
1885 inventions and musical instruments
1886 products of the colonies and India.

This was followed by a further series of exhibitions, also to be held in London, featuring the products of individual countries. They took place in a new venue at Earls Court, Kensington, completed in 1887 especially for public exhibitions of this kind, and were [15]:

1887 America
1888 Italy
1889 France
1890 Germany.

Some of these exhibitions contained electrical artifacts. The American exhibition included *'apparatus for the production and application of electricity'*, and a few electrical devices were found in the Italian and German exhibitions, but nothing that had not been seen in earlier exhibitions. The French displayed an electric apparatus *'for registering the revolution of a wheel in mechanical devices'* and displayed a selection of *'carbons for electric piles, (arc)lights and microphones'*.

At the Inventions and Musical Instruments Exhibition in 1885, Sir Charles Parsons was able to show for the first time in public his turbine driving an alternating current generator. He patented this in 1884, and initiated a technique which was to dominate the electrical industry in the following years. The earlier generators had been limited by the low speed of the reciprocating engines used to drive them, and conversion to the use of the much higher speed steam turbines increased the efficiency of AC generation. It was the Anglo–Brush company that exhibited a double-flow turbine, based on a Kitt–Parsons design, which operated at 12 000 rev/min and was coupled to a small alternator generating a current of 75 A at 100 V.

Sir Ambrose Fleming, a frequent visitor to the Exhibition, describes an AC transmission system, shown by a Hungarian engineer M. Zipernowsky, and brought over from the Hungarian National Exhibition. This demonstrated how, with the aid of a transformer, the output of an alternator could be conveyed long distances at a high voltage with small energy loss. Several such transformers were used with their primary circuits arranged in parallel on the alternator, each supplying a number of lamps from the secondary winding. It was found possible to control the lamps on any one transformer without affecting those on any other [16]. A similar technology had been shown a year earlier in 1884 at the Turin Exhibition but with series connections of the transformers across the alternator which resulted in a lower efficiency (see Chap. 7).

A major feature of the Inventions Exhibition was its extremely large collection of working dynamos in the Electric Light Machinery Shed. There were 72 of these, varying in capacity, to feed lamps ranging from a few hundred candle power to tens of thousands. Each of the 35 exhibitors set up its own generators to drive a set of arc and incandescent lamps, located at an exhibition stand or a specific area such as a dining room, entrance or annexe. This meant that not only did the display of generators cover a wide range of types and manufacture but almost all the known types of lamps developed in the previous decade were seen together, albeit in a haphazard manner [17]. The prime movers, steam or gas engines were, however, often shared between exhibitors. Similar but rather better organised collections of generating equipment and lamps were to be seen in London and in several other locations in Britain during the 'Jubilee' exhibitions of 1886–88. However, since a major influence for these exhibitions was the early development of electrical power generation these are considered separately in Chap. 8.

None of these London exhibitions were 'international' in the sense that the 1851 and 1862 exhibitions had been, although the fisheries exhibition in 1883 received a considerable input from the United States. In general, however, they included few items from outside Great Britain and were not open to receive contributions from all on a competitive or comparative basis.

One small but highly successful public international exhibition, taking place in 1876 at the South Kensington Museum, London, was the 'Special Loan Collection of Scientific Apparatus'. This was opened by the Queen and provided a focus of attention for science and engineering at the time. Sir William Preece, then President of the Institution of Electrical Engineers, writing in 1882, described the Loan Collection of 1876 as, ' ... The first real electrical exhibition' [18].

It consisted of an amalgam of artifacts provided by the Kensington Museum, the Patent Office collection and items from abroad, some of which are shown in Figure 6.3. Amongst them may be seen Sömmering's electrochemical telegraph. This is often claimed as the first electrical telegraph, although pithball electrometers and other devices, utilising static electricity, were in use toward the end of the 18th century. Samuel Thomas von Sömmering demonstrated his telegraph transmitter/receiver in Munich in 1809. It consisted of a number of circuits with a common return wire, each capable of

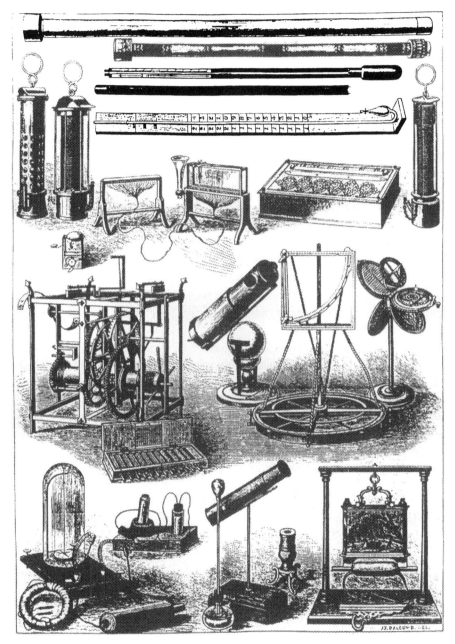

Figure 6.3 Some items shown at the Special Loan Collection of 1876
(Illustrated London News Picture Library)

carrying a small current to an electrode situated in a glass tank full of
acidulated water (the receiver). Closing a given circuit in the set of
transmitter switches initiated the process of electrolysis detected by

bubbles of gas rising to the surface of the tank from the energised electrode [19]. One important experimental demonstration shown at this exhibition and considered in more detail later, was the transmission of mechanical power from one dynamo machine to another, acting as a motor, through a length of conducting cable [20]. This was by M. Gramme, the Paris engineer who had originally demonstrated this far-reaching technology at the 1873 Vienna exhibition.

The Philosophical Society of Glasgow held a small technological exhibition in 1880 on 'Apparatus for the utilisation of gas, electricity etc', with a comprehensive sector on electrical applications. These included Mackenzie patent electrical lamps, 'worked by a Gramme dynamo-electric machine', four Serrin–Lenton arc lamps and a number of Brush and Crompton lamps. One of the Serrin–Lenton lamps was situated on the top of the University building where an occulating light sent out a Morse code dot and dash every 15 s. A telephone exchange system with a telephone, '... of excellent construction and a loud sound', a street fire alarm, a magnetic fuse exploder and indicators and batteries for use in ships were demonstrated [21].

Although Britain provided a number of other exhibitions with a technical content in London and the provinces during the second half of the 19th century [22] (see Table 2.1), and contributed quite effectively to international exhibitions abroad, none were outstanding, and the impetus for arranging progressive and exciting international collections now passed to continental Europe, specifically France and Germany and, towards the end of the 19th century, to the United States.

6.5 References

1 ROLT, L.T.C.: 'Victorian engineering' (Allen Lane, London, 1970), pp. 148–177
2 FIGUIER, L.: 'Les merveilles de l'industrie' (Corbeil, Paris, 1862)
3 CLARK, D.K.: 'The exhibited machinery of 1862, a cyclopaedia of the machinery represented at the International Exhibition' (Day & Son, London, 1864)
4 VON WEIHER, S. and GOETZELER, H.: 'The Siemens Company—its historic role in the progress of electrical engineering' (Franz Steiner Verlag, Wiesbaden, 1977), pp. 12–13
5 TIMBS, J.: 'The industry, science and art of the age—The International Exhibition of 1862' (Lockwood, London, 1863), pp. 141–142
6 MALLETT, R. (Ed.): 'The Record of the International 1862 Exhibition' (Wm Mackenzie, London, 1862), pp. 531–533
7 'Lighthouses', *in* 'Encyclopaedia Britannica, Vol. 30' (Encyclopaedia Brittanica Inc., Chicago, 1923) pp. 257–260
8 SINGER, HOLMYARD, HALL and WILLIAMS: 'A history of technology, Vol. V, 1850–1900' (Oxford University Press, 1958), p. 181

9 ASH, E.A.: 'Education for an age of technology', *in* SINGER, HOLMYARD, HALL and WILLIAMS: 'A history of technology, Vol. 5' (Oxford University Press, 1958), Chap. 32, pp. 776–780

10 ATKINSON, J.B.: *Blackwoods Mag.*, April 1862, **XCI**, p. 476

11 WILDMAN, S.: 'Great, greater, greatest? Anglo-French rivalry at the Great Exhibitions of 1851, 1855 and 1862', *J. RSA*, 1989, pp. 137, 660–664

12 Report, 'Plans for a London Exhibition of 1871', *J. RSA*, 6 August 1869, **17**, p. 13

13 HILL, G.P., and WRIGHT, J.C.: 'British history 1815–1914' (Oxford University Press, 1981)

14 ELLIS, A.J.: 'Sensation of tone' (translated from Helmholtz) (London, 1885)

15 LOWE, C.: 'Four national exhibitions in London and their organisation' (Fisher-Unwin, London, 1892)

16 FLEMING, J.A.: 'Fifty years of electricity' (The Wireless Press, London, 1921), p. 133

17 'The international inventions exhibition', *The Electrician*, 9 May 1885, **18**, pp. 1–14

18 PREECE, W.: 'Electrical exhibitions', *J. RSA*, 1882, **31**

19 FLEMING, J.A.: see 'The electrical exhibition in Paris', *The Electrician*, 15 Oct. 1881, **8**, p. 341

20 'Gramme's demonstration of the transmission of electric power', *Engineering*, 1889, **28**, p. 417

21 'Reports relative to an exhibition of apparatus for the utilization of gas, electricity etc.'. Philos. Soc. Glasgow (C. Anderson, Glasgow, Sept/Oct 1882)

22 'Memorandum on exhibitions held in Great Britain and Ireland'. Pamphlet no. 1889, Royal Society of Arts, 1903

Chapter 7
International exhibitions abroad

7.1 New York, 1853

The highly successful 1851 exhibition in London encouraged much activity abroad during the remainder of the century in establishing similar events. In the United States the first attempt in 1853 to create an international exhibition was however a disaster. A group of New York City businessmen, including P.T. Barnum, of Barnum's circus, and H. Gresley, editor of the *New York Tribune*, planned together an international exhibition on a 5.26 ha site in the heart of New York City, with its main architectural feature a building modelled on the British Crystal Palace. The design of the building was, however, poor; it leaked badly, often ruining the exhibits and soaking the visitors! A severe financial loss ensued.

It did, however, contain a number of electrical exhibits which made the Exhibition of interest to technologists [1]. J.B. Richards of New York demonstrated House's printing telegraph between offices situated at opposite ends of the building. W.M. Swain, the president of the Morse Magnetic Telegraph Co., showed *'Morse's patent electric telegraph in operation with wires in direct connection with all the principle lines in the United States'*. J. Isenring displayed an electromagnetic conductor and storm indicator, which apparatus *'is put in connection with a lightning rod, and will indicate a storm at 307 hours in advance'*, presumably by slight discharges observed in advance of the main storm.

Various magnetoelectric machines were shown and dismissed by Professor Maurice Vergnes as, *'...but none have gotten over the general objections which apply to all, and among which the great expense is prominent'*. Electricity still retained its mystery and, amongst the display of magnetoelectric machines for medical purposes, was displayed

'*...an ingenious modification of the common voltaic pile, consisting of a chain composed of a series of zinc and copper wires arranged on a piece of porous wood; by immersing the chain in vinegar for a short time the porous material absorbs sufficient fluid to keep it active half an hour. A chain of 60 metallic elements may be joined together; the chain being perfectly flexible and may be worn a long time under the garments and its mild but long continued influence is sufficient to obtain results. These are continuous or interrupted, accompanied by shocks through a very simple mechanical contrivance. It may be used for galvano-puncture; and ... is sufficiently powerful for the decomposition of water and metallic salts in solution .... The testimony of the most eminent surgeons in this country and in Europe goes to prove that it is an efficacious, portable and economic instrument for medical purposes*'. ,

A similar exhibit had been shown on this side of the Atlantic at the 1851 Exhibition (see Chap. 5).

Other exhibits included the first American showing of Elias Howe's invention of the sewing machine and a very large display of American agricultural machines which had caused so much interest in Britain when shown at the Great Exhibition of 1851.

7.2 Dublin exhibitions of 1853 and 1865

Later that same year the Dublin World Fair was opened. The organisers of the Dublin exhibition had showed similar architectural ideas to those of their counterparts in New York and erected a temporary building, having some resemblance to the London Crystal Palace, on the capacious lawns of Leinster House, Dublin. Unlike Paxton's design it contained little glass and substituted wooden panels for most of the cladding of a large circular dome, the main feature of the structure. The building was certainly more successful than the New York venture. The Irish had considerable earlier experience in arranging national exhibitions for manufacturers (this was their eighth), from as early as 1826 with succeeding exhibitions in Dublin in 1829, 1834, 1837, 1845, 1850, and 1852 (in Cork). This last was moderately successful; it made a slight surplus, attracted 140 000 visitors and encouraged Dublin to plan its own, larger, exhibition.

Their first international exhibition in 1853, known as 'The Great International Exhibition of all Nations' was intended to *display the materials of manufacturing arts, with a view to practical results in the development of national industries'.* It was in fact an industrial-orientated exhibition conceived as

> *'a centre whence a stream of new industrial life would flow back into every valley and over every plain'.* The Times of the 23rd May 1853 echoed these sentiments and hoped, *'... it would end Irish migration to the already overcrowded labour markets of England'.*

This was not to be the case. Efforts to secure exhibits from abroad met with little success, with many of the potential contributors in Europe preferring to send their exhibits to the rival Crystal Palace in New York. Great Britain was a major contributor with 1500 exhibits, some from the 1851 exhibition, but only 254 came from other countries overseas. This left the organisers to fill up the many empty spaces with additional exhibits from Ireland itself, and this now included a large contingent of arts exhibits. As described by Davis [2],

> *'machinery and manufacturers (were) relegated to the sidelines, obscured by paintings, sculpture, exotic oriental and luxury European exhibits'.*

A certain number of industrial exhibits remained. An impressive hall termed 'machinery in motion', contained two 25 hp high-pressure steam engines, made by Fairbairn of Manchester, which transmitted power by means of a shaft running along the length of the hall to drive a number of machines by the use of belt drives. These included a printing press, which daily produced the exhibition's newspaper, a power loom, spinning machines and a flour mill. Other 'philosophical and surgical items' contained a Grubb telescope, some telegraphic machines, daguerreotypes and photographs of the 1851 London exhibition. Various branches of engineering sent exhibits, and outside in the grounds, agricultural machines and over 60 examples of carriages were to be seen.

Towards the end of its six months' duration over one million visitors had seen the exhibition, which provided one legacy in an unexpected direction, namely the erection of a public gallery of art, which eventually became the Dublin National Gallery [3, 4]. There is no record of any useful assistance, however, to industrial development in Ireland arising from the exhibition.

Dublin was to open a second International Exhibition in 1865. The main building, the Palace and Winter Gardens Building, was erected in the grounds of Coburg gardens, again using Paxton's iron and glass design, seen in Figure 7.1 during the course of erection, and

Figure 7.1   Construction of the Palace Building for the Dublin 1865 Exhibition (Parkinson, 'The Dublin International Exhibition of 1865')

augmented by a machinery annexe. The building had a floor area of 4770 m², and contained three floors, a basement and a gallery. The classification of the exhibits was the same as in the 1853 exhibition and similar to that adopted by the London 1851 exhibition.

The electrical exhibits, housed in section VII—civil engineering, were almost entirely confined to a significant contribution by the Siemens company, which was anxious to extend its marketing to cover Ireland from its newly established company in Britain. The telegraphy exhibits included some telegraph recording instruments, a magneto alphabetical telegraph with alarm, a railway alarm, testing instruments, induction coils, galvanometers, telegraph posts, etc., and various working telegraph apparatus including a Morse ink recorder. The exhibits also showed their 'hotel telegraph equipment', '... similar to that installed in the Hotel de Louvre and Grand Hotel in Paris, and Charing Cross Hotel in London'. A single bell was used and the room indicated by a relay which acted to uncover a number appearing in a glass window. Siemens showed a range of land and submarine cables, electrical equipment for mines and a mine exploding relay [5].

7.3 Four French international expositions, 1855, 1862, 1867 and 1878

The two French international expositions of 1855 and 1862, which followed the Great Exhibition, did not contain any major technical innovations. Perhaps the interval was too short to allow industry to perfect any new ideas that would have been present in 1851. The 1855 exposition was, however, significant in one engineering achievement of some importance. This was the first public exhibition displaying the newly discovered material, aluminium. It proved at first so difficult to refine that it was classed as a precious metal and Napoleon ordered a dinner service to be made of it after a visit to the exposition. Later the metal was produced by an electrolytic process developed independently by Hull in America and Heroult in France. For many years afterwards Switzerland, with its abundant water resources used for electricity power generation, was the largest European producer.

In general the expositions showed a similar but smaller set of exhibits of the kind seen at the British 1851 Exhibition. One new venture was a display entitled, 'l'exposition economique' which exhibited cheap manufactured articles of practical use and which led, in subsequent expositions, to whole exhibition areas being concerned with domestic products.

The largest building at the 1855 exposition was the Palais de l'Industrie, a large massive edifice in the heavy style of the period (Figure 7.2), far removed from the lightness and spectacle of the Crystal Palace. In an annexe to the building was a steel framed structure, the Galerie des Machines, containing a number of locomotives and other steam machines. A separate section elsewhere in the exposition paid homage to men of science with details of the work of Scheele, Priestley, Cavendish, Lavoisier, Richter, Volta, Dalton, Davy, Wollaston, Berthallet, Bergman, Berzelius and Gay-lussac [6]. The exposition attracted over five million visitors. Queen Victoria with Prince Albert made their first visit to France to see the exposition (the first visit by a reigning British monarch since Henry V). The 1855 and 1862 French exhibitions both included an extensive and magnificent art display, which encouraged the British to extend their exhibitions in a similar way from 1862 onwards to the detriment of the significant technical and engineering achievements the British were capable of displaying at that time.

The Paris expositions of 1867 and 1878 also included sizable art displays, but in addition they gave much more attention to interesting and educating the general public in a way not previously attempted in the earlier 'trade show' activities in France prior to 1867. A new feature at the 1867 exposition was a section devoted to 'the history of labour' with a class on 'social and moral problems' [7]. This marked the beginning of a tendency for international exhibitions to develop increasingly along cultural rather than commercial lines, a trend referred to later in this book. The main building for this and later exhibitions in Paris, the 'Palais du Champ de Mars', was commissioned by Frederic le Play and also modelled on the 1851 Crystal Palace. It consisted of a great elliptical iron and glass structure covering 50 ha with raised galleries for spectators and referred to as 'la merveille du genre'. It was the first exhibition building of its type to have the stresses and strains of its structure properly computed before erection by its designer, Gustave Eiffel, who was to be associated with the architecture of public exhibitions for a number of years. The 1867 exposition contained 42 000 exhibits, including for the first time in Europe a sizable contribution from the United States.

The French had been notable for their decided views on classification of exhibition products, causing not a little difficulty in discussions with the organisers of the 1851 Great Exhibition. At the 1867 exposition the French planners were able to put their own ideas into effect. Since the 1851 London exhibition the accepted arrangement for international exhibitions was to partition the exhibition

Figure 7.2 The Palais de l'Industrie at the Paris 1855 Exposition (L'Illustration Journal Universelle, 1854)

space into separate areas, one for each of the contributing nations, and to allow them a free hand to arrange their exhibits as they pleased. In the Palais du Champ de Mars all the contributing nations were asked to arrange their exhibits into allied groups so that similar items could be displayed against similar artifacts from competing nations. This was achieved by using a ground plan consisting of a series of concentric elliptical rings, increasing in length from the centre outwards, as may be seen from Figure 7.3, with each ring allocated to a particular class of goods and each sector to a different country. For example, the outermost and largest ring was devoted to machinery which, while suitable for a highly industrialised country such as Great Britain having much to display, was not appreciated by an underdeveloped country which had little machinery to exhibit. In general this new form of classification worked reasonably well and made it easier for the visitor to follow the progress of particular technologies, comparing one nation's effort with another. It was certainly helpful in displaying the devices from a fast developing technology, such as the emerging electrical industry, and particularly for telegraphy.

By 1867 the Chappe system of semaphore telegraphy in France had given way to the Morse system, and a large number of such systems

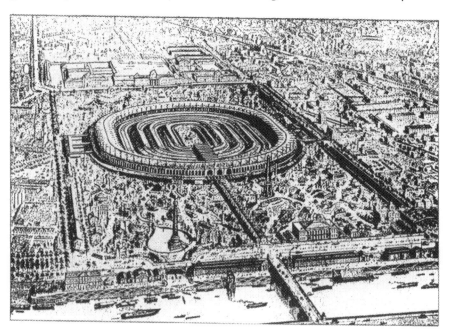

Figure 7.3   The plan of the Paris 1867 Exposition
(Official Catalogue of the 1867 Exposition)

were shown by French and foreign exhibitors. Many included automatic recording and printing apparatus, typically those of Meyer, Joly, Hardy, Favrel, Sortail, Guillot, Bréguet (a major French inventor and manufacturer), Siemens & Halske, and Bonelli from Italy. The military made an appearance at an exhibition for probably the first time in any numbers, and showed several army telegraph systems, including one with the imposing title of 'Galvanic Establishment of the Corps of Engineering of St. Petersburg, Russia'. From Britain, HM Secretary of State for War displayed some of Wheatstone's military electric telegraphs. From Germany came an apparatus for measuring the trajectory of a missile by Captain Schultz. The difficulties of laying and recovering submarine cables were beginning to exercise the minds of many manufacturers, and several companies showed models and exhibits relating to this activity. The Morse company from the United States was foremost in this as well as Balestrini of Italy and Reuters of London, with the last named exhibiting specimens of the recent (1866) submarine cable laid between England and Hannover.

The American contribution to the 1867 Paris Exposition was committed only a few weeks after the end of the Civil War, despite a *'staggering war debt and an unstable financial policy ...'* [8]. Nevertheless the American contributions won many prizes. A Grand Prize was awarded to Cyrus W. Field for his recent work on the transatlantic cable, featured at the exhibition, and a gold medal awarded for a telegraph printing machine. America was a major contender in the field of telegraphy at that time due to its importance for both sides in the civil war. The first field line telegraph was established in Virginia in July 1861 between McClellan's headquarters and its front line officers. Some 8694 km of telegraph line was constructed in the northern states between 1861 and 1866, more than 1610 km by the military [9]. The first exhibition abroad of the new Corliss beam engine was shown in the American section. This was a smaller version of the mammoth 1500 hp machine, which was to be seen at the Philadelphia exhibition in 1876 (Section 7.4).

Léclanché exhibited a number of batteries and accumulators at a time just prior to his important development of the dry cell in 1868 which revolutionised portable low-power supply (100 years later the world market for such cells was to exceed more than 500 million units [10]). Another well known name in electricity, Rühmkorff, displayed electric and magnetic instruments.

The Siemens company showed their AC dynamo, one of the first to make use of residual magnetism for start-up. This was displayed as a static exhibit since no power was available to drive it. The generator was to find a practical role in the following year when it was used by the military both as a source of power for electric arc-lighting for bridge-building at night, and as a method of detonating explosives for mining operations [11]. The exposition saw the first appearance of the ball-bearing, a British invention, which was to have such a significant effect on the efficiency of moving machinery.

Although Britain contributed over 3000 exhibits to the Paris 1867 exposition these were not viewed at all well by influential visitors. In a letter to the RSA, contributed by Dr Lyon Playfair, who we noted earlier was responsible for the classification of material for the 1851 exhibition, he writes,

> '...I am sorry to say that, with very few exceptions, a singular accordance of opinion prevailed (in Paris and amongst other European visitors) that our country has shown little inventiveness and made little progress in the peaceful arts of industry since 1862', and puts this down to '... the lack of good systems of industrial education for masters and managers of factories and workshops' [12].

Dr Playfair was one of the first to recognise the vital role of science and technology in education and devoted much effort to bring about the necessary educational reforms in Britain which followed the 1862 and 1867 exhibitions.

The 1878 exposition, also in the Champ de Mars, was the first opportunity for the Third Republic to demonstrate its industrial prowess, following the disasters of the Franco–Prussian war of 1870–1871 and the Commune Revolt. Although no longer extolling Royalty and the Empire and instead fêting the Republic, it was first and foremost a technology exposition. It was even bigger than that of 1867 with additional sites at Ville de Paris and Quai d'Orsay [13], with prominence given to the new developments of Bell's telephone, Hughes' microphone and the phonograph system of Edison, this last now in the Science Museum, South Kensington. The latest version of Hughes' printing telegraph was also exhibited (Figure 7.4). This made use of a piano keyboard having 14 white and 14 black keys, sufficient to allocate a letter of the alphabet to each key. It was first seen in Europe at a private models' exhibition of the RSA in 1860. In the intervening period it had almost supplanted all other such devices, owing to its simple but effective synchronisation between transmitter and receiver mechanisms, a problem that had plagued earlier designs [14].

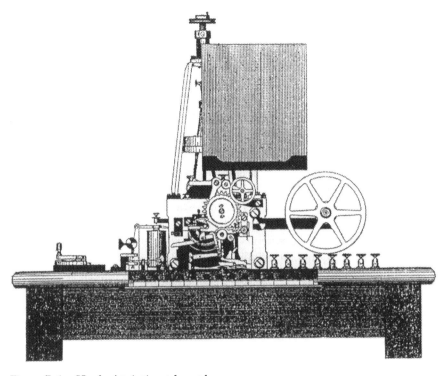

Figure 7.4   Hughes' printing telegraph
(Fleming, 'Fifty years of electricity', 1921)

An extensive display of telegraph apparatus was shown by the French containing all the range of equipment seen at previous international exhibitions. The finish of the French manufacturers' designs was outstanding, prompting the *Illustrated London News* to comment in one of their special issues, brought out for the Exposition,

> '... *their form of apparatus are characterized by a neatness and artistic taste, which are rare amongst English makers, and are perhaps only equalled by those of America ... and (regarding the telephone equipment) the English equipment is turned out of common mahogany and varnished, so that it has a cheap appearance ... in France it is an extremely handsome little instrument, sometimes nickel or silver-plated*' [15]!

The French telephone instrument was sold at the Exposition for 24 F and 30 F with the visitor advised to, 'buy one for his own use and take it back to England'.

In the special acoustics section of the exposition an experimental telephone system was set up in which

> '... *Messrs Garnier and Pollard have contrived to combine Mr Edison's sending apparatus with Mr Bell's telephone, but replacing the permanent magnet of the latter with an electromagnet. A conducting wire from a pile of 10 elements Léclanché was connected in its circuit for the two instruments.*'

This arrangement had been shown earlier at the Academie des Sciences in Paris by the industrious M. Bréguet. The microphone referred to as a 'sending apparatus' was an original discovery of Professor Hughes (and exhibited as such in the Exposition). The Hughes microphone shown at the exhibition used iron filings and, when introducing this to a meeting of the Royal Society, Professor Hughes remarked that,

> '... *I do not intend to take out a patent, as the facts ... belong to the domain of discovery ... no doubt inventors will before long improve on the form and materials I have used*'.

It was later improved through the use of carbon granules and patented by Edison in 1877.

An early display of multiple telegraph instruments was shown with Mayer's instrument demonstrating the transmission of up to eight Morse messages simultaneously along a single line. The wires laid for telegraphic purposes represented the largest capital outlay in a telegraph system so that multiplex working was important in reducing the installation costs for the telegraph companies. In the British section Varley showed his magnetoelectric machine, electric motors and a needle telegraph. Another British item was Andrew Jamieson's patent self-relieving grapnel for raising submarine cables, a technique in which the British excelled at that time.

As well as contributing much to the 1878 exposition, Edison was also to dominate the display of lighting in the city. In June he switched on for the first time public lighting that had been installed along the Avenue de l'Opera and the Place de l'Opera at a time to coincide with the 1878 exposition. These were carbon-arc installations using Jablochkoff candles. They had only recently been invented by Pavel Jablochkoff, a Russian telegraph engineer then resident in Paris, and consisted of two parallel carbon rods separated by a thin layer of plaster of Paris. A thin piece of graphite joins the carbon rods at their upper end. When the current is switched on this graphite piece is vaporised and an arc is struck which remains across the top of the carbon rods as the rods burn and the plaster layer crumbles away. Owing to the nature of their construction they could be used on either AC or DC, although there were difficulties of uneven consumption of carbon in the two rods if DC was used. The installation was quite extensive, consisting of about 400 lamps, requiring three DC generator stations along the route, each driven by a 16 hp steam engine [16]. They lit the thoroughfare with greater brilliance than had ever been possible with gas lighting, causing a commentator to exclaim that, *'...whatever may be the future of the electric light, its extension will be traced to the Paris Exposition'* [16]. The Jablochkoff candles were not destined to remain in the Paris streets for very long, however, being superseded by reliable regulated arc lamps as soon as these became available commercially. The main problem with the candles was that they needed replacement every time the light was switched off, although this was obviated to some extent by arranging the candles in groups of four and a fresh candle automatically cut into the circuit as each preceding candle was consumed [17]. Their replacement was another job for the street lamplighter (Figure 7.5). In their short life in the Paris streets, however, they certainly served to bring the advantages of electric lighting to the public and to encourage its further development.

The Exposition had two narrow machinery halls, each 653 m long and 35 m wide in which the exhibitors were assisted by having, *'Water, gas and steam ... supplied gratuitously but all countershafting, intermediate gearing and belting provided by exhibitors'*. Approximately 1200 hp was necessary to drive the whole of the machinery in motion [18]. One of the halls was used by the French and the other shared by foreign exhibitors from England, the United States, Belgium, Switzerland and Austria.

A Royal Commission from England visited the 1878 exposition [19] and were impressed by its size and technical content—so much so that on their return a letter was sent to the Council of the RSA with a somewhat unusual request. They were asked to undertake the duty of sending a number of selected artisans to Paris and for each of them to write a report on his own speciality, as seen at the Exposition and in the workshops of Paris. This produced a valuable set of reports giving a sound view of craft and artisan work in 19th century France [20]. Two hundred and four artisans, selected with the assistance of the mechanics' institutes, were sent from towns all over Britain. They covered almost all the industries working at the time, but

Figure 7.5   Replacement of carbons in the Jablochkoff candle
(Illustrated London News, 1878)

unfortunately no one was selected who was knowledgeable about the small-scale industries covering the cable, telegraph and electrical fields. Each artisan was furnished with a return rail pass from London to Paris (£1), £8 pocket money and advice as to where to secure 'suitable lodgings in the neighbourhood of the Exhibition at a maximum charge of 20 F (16 shillings) *per week!*' and 'good dinners, served as in England, at 50 cents (1/3d) at the Workmans Hall, near the Exhibition'.

The Exposition of 1878 admitted 16 million visitors, almost three times the number attending the 1851 Great Exhibition, and clearly established France's lead in this field—but not for long. The entire character of international exhibitions was due to change as a result of the swift advance in technology, particularly electrical technology, following the inventions of Gramme, Swan, Edison and others. This lead was soon to be challenged first by Germany and, as the century progressed, by the United States.

7.4 The Philadelphia Centennial Exhibition of 1876

The industrial strength of the United States was beginning to be felt, however, before 1878 in the first successful United States international exhibition, the large Philadelphia Centennial Exhibition of 1876 [21]. This Exhibition was arranged to mark the centenary of American Independence and claimed to '... *foster trade, stimulate economic growth ... diffuse scientific and technological information etc'*. The Franklin Institute played a major role in its organisation having already held an earlier exhibition of its own in Philadelphia in 1874.

The location chosen for the 1876 Exhibition was Fairmount Park in Philadelphia. This had a total area of over 1215 ha, of which the international exhibition covered about 182 ha including an extensive fairground. Two major buildings were featured: the main building devoted to manufacturers in the United States and other countries (37 nations took part in the exhibition); and a huge machinery hall. A machinery hall had now become a feature of all international exhibitions, initially for the display of large mechanical engines, and later to house electrical generating machinery required to power the various lights, motors and appliances throughout the exhibition site. Since the date of the Exhibition was just prior to a satisfactory industrial dynamo becoming available, it was dominated by heavy mechanical engines, including the largest twin Corliss beam engine then exhibited (1500 hp) driving all the 3400 m of shafting in the machinery hall (Figure 7.6). The engine was started by President Grant on opening the exhibition and ran continuously for six months, the duration of the Exhibition [22].

Most exhibits in these halls and others throughout the site were grouped according to country of origin, departing from the system applied in the French expositions. A further classification was then

Figure 7.6   1500 hp twin Corliss beam engine, Philadelphia Exhibition, 1876
(Norton, 'Register of the Centennial Exposition, 1876')

made according to eight departments: mining, metallurgy, manufacturing, education, science, art, machinery, agriculture and horticulture; and then further subclassified and subclassified again in a logical manner, that later became a model for the Dewey Decimal System, now widely used in libraries throughout the world. Three thousand exhibits were sent from Great Britain, the largest foreign contribution from any country exhibiting.

In the machinery hall was shown, for the first time, a practical typewriter—a mechanical wonder at the time. Also shown was the Grant calculating machine—a mechanical difference engine used in the construction of *'large mathematical tables eg logarithms, sines, tangents, reciprocals, square and cube roots'.* It was built for the University of Pennsylvania and its descriptive material states, *'...those interested will be familiar with the Difference Engine of the late Charles Babbage and its failure ...'.* It was apparently of massive construction, 1.5 m wide and 2.5 m long and weighing 908 kg, *'with a front end apparatus for printing a wax mould of the results from which an electrotype may be made'.*

The electrical exhibits were not very extensive but did provide a good display of telegraph equipment, including a working 'signal station' manned by an army chief signals officer and containing '... telegraph and barometric instruments all of American invention' (mainly by the Western Electric Manufacturing Co. of Chicago). An Anders magneto printing telegraph was shown which was worked 'without any batteries', with electricity generated through an armature rotating between the poles of a large magnet using a treadle borrowed from a sewing machine. As used in the Hughes printing telegraph, described earlier, a piano keyboard was employed in which 26 black and white keys actuated relays to send out signals associated with each letter and, at the same time, to use these to initiate a printing mechanism to record each letter on a strip of paper carried on a spool (Figure 7.7). Edison demonstrated some of his experimental incandescent lamps, which were at this time rather fragile and not yet suitable for commercial manufacture.

The Gramme company gave a demonstration of the electrical transmission of power, as seen earlier at the Vienna Exhibition of 1873, where two identical machines were used situated some way apart, one as a generator and the other as a motor driving a pump (see also the following Chapter).

An important item at this Exhibition was a first display of Graham Bell's telephone, at that time a simple membrane transmitter and receiver, contained in a wooden box, with which the talker had to alternately talk and listen. This is shown in Figure 7.8 with acknowledgments to the Science Museum, London, where the original is kept. Its range was limited since its only source of power was the human voice. At first this exhibit went unnoticed by the visitors, on account of its location in an obscure corner of the Education Building, away from the other exhibits. Then came the day of the judge's

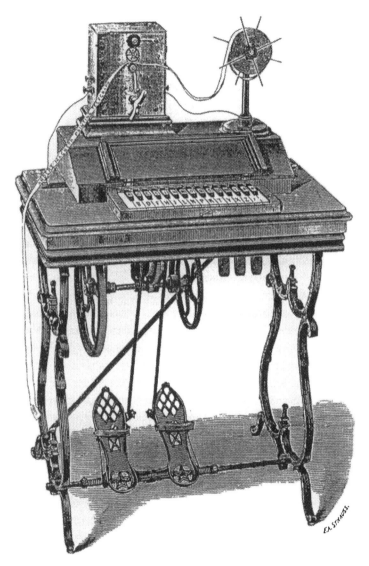

Figure 7.7   Anders magneto printing telegraph
(Norton, 'Register of the Centennial Exposition, 1876'

inspection of exhibits for prizes to be awarded. They duly looked at the rather odd looking exhibit and were prepared to pass on when Don Pedro, the young Emperor of Brazil, who had some prior knowledge of Bell's teaching abilities with deaf children, placed the receiver to his ear whilst Bell went to the other end of the line and spoke to him. In amazement the Emperor exclaimed, *'My God—it talks!'* Grouped about the Emperor were a number of spectators

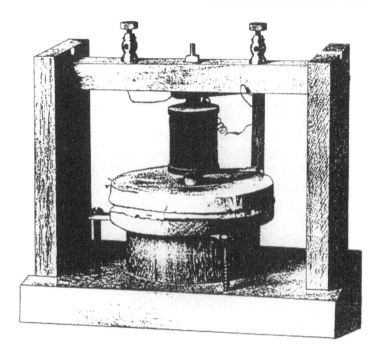

Figure 7.8   Graham Bell's telephone, first shown at the Philadelphia Centennial Exhibition of 1876
(Science Museum Photograph)

including two distinguished scientists, Joseph Henry and Lord Kelvin. After each in turn had listened to the voices coming over the wire, they announced to the Judge, *'Here is the greatest marvel ever achieved in electrical science!'* Needless to say Bell was awarded one of the Exhibition's coveted prizes for his telephone and also for his 'harmonic telegraph', described earlier in Chapter 6. Bell also exhibited his liquid resistance telephone in which a metal rod was attached to the diaphragm and inserted into a liquid forming part of a battery circuit. The vibrations of the diaphragm in the liquid varied the electrical resistance of the circuit and hence provided a variable current corresponding to the sound waves falling on the diaphragm. This invention did not, however, attract as much interest as his magneto type of design. Sir William Preece, who was visiting the exhibition at that time, obtained a prototype telephone made by Graham Bell, and brought this back to England. It was shown by him at the BAAS meeting in Glasgow in the September of that year, the first example of a telephone seen in this country.

With the electric exhibits could also be seen a number of applications which were new at the time. These were burglar alarms, thermostatically controlled relays, and a rather fearsome 'dentist's electromagnetic mallet', which could also be used for 'light engraving on metal'! Several well known, but still intriguing to the public, pieces of 'family electromedical apparatus' demonstrated their power to shock the unwary visitor.

The British contribution to this exhibition was minimal since the industry was unwilling to exhibit in a country protected by high tariffs (30–40% on foreign goods). Other countries took part, however, and in particular the Japanese, who not only contributed to the exhibits but also arranged a visit of 66 of their technical engineers. On their return to Japan, they produced a monumental 96 volume report, effectively providing Japan with much of the necessary information to set her on her way to becoming a fully industrial nation following the advent of the Meiji period of modern government in 1870.

7.5 Other international exhibitions abroad, 1870–1885

The years 1870–1885 were very productive of exhibitions, as may be seen from Figure 7.9, which shows a rolling average of the yearly number of technological exhibitions between 1800 and 1990. Some of these were devoted entirely to the electrical industry and will be considered in a later Chapter. A fairly complete list is shown in Table 2.1.

The new advances in technology were seen in many of the minor exhibitions held in widely divergent locations. In Russia the Polytechnic Exhibition in Moscow celebrated the bicentennial of the birth of Peter the Great. It included separate halls devoted to geology, mineralogy, mining, technology, applied physics, botany and horticulture. The exhibits were later used to form the present Central Polytechnic Museum. At the Turin Exhibition of 1884 an extensive

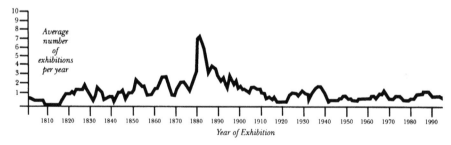

Figure 7.9   Rolling average of the yearly number of technological exhibitions held between 1800 and 1990

demonstration of the Gaulard and Gibbs electrical transmission system, first seen at the Westminster Aquarium Exhibition in 1883, was shown. A transmission line was built between Lanzo and Turin, a distance of 40 km, with illumination provided at both ends and along the route [23]. The line consisted of chrome-bronze wire strung along ordinary telegraph posts and achieving a conductivity of 98%. This was supplied by a Siemens generator requiring a 30 hp engine to drive it [24]. The success of this project won a premium of 15000 F for '... *The best means of transmitting electrical energy to great distances*' [25]. It was this demonstration at Turin which prompted George Westinghouse to purchase the American patents for the system and so introduce an AC distribution system into the United States for the first time in 1885–86 [26]. An unexpected demonstration of the effects of electricity occurred during the Turin Exhibition when a great captive balloon, forming one of the attractions, was struck by lightning which ignited the 1000 m^3 of gas contained within it and so provided a spectacular display to the citizens of Turin!

Other early exhibitions with an electrical content include one at Antwerp, at which in 1885 an international display of mechanical and electric traction that featured all types of motor vehicles existing at that time was presented. Long-distance telecommunication demonstrated at this exhibition was inaugurated between Brussels and the main cities of the country, enabling Queen Marie Henrietta during her visit to listen to a performance of 'Faust' relayed from a Brussels theatre.

A major influence affecting exhibitions towards the end of this period was, however, the availability of a sufficient quantity of electric power to make exhibition lighting not only feasible but spectacular, and this is the subject for the next Chapter.

7.6 References

1 GOODRICH, S.R.: 'Science and mechanism—New York Exhibition 1853–4' (G.P. Putnam & Co., New York, 1853)
2 DAVIES, A.C.: 'Ireland's Crystal Palace, 1853' GOLDSTROM, J.M., and CLARKSON, L.A. (Eds.): 'Irish population, economy, and society' (Clarendon Press, Oxford, 1981)
3 JONES, T.D.: 'Record of the Great Industrial Exhibition (Dublin 1853)' (Dublin, 1854)
4 BATTERSBY, W.J.: 'The glories of the Great International Exhibition of all Nations in 1853' (London, 1853)
5 PARKINSON, H., and SIMMONDS, P.L.: 'Dublin International Exhibition 1865' (London, 1866)
6 'Visit of Prince Napoleon to the Palace of Industry and Guide to the Exposition of 1855' (Perrotin, Paris, 1855)
7 LUCKHURST, K.W.: 'The story of exhibitions' (Studio Publ., London, 1951), p. 132

8 OLIVER, J.W.: 'History of American technology' (Ronald Press Co., New York, 1956), p. 297
9 PLUM, W.R.: 'Military telegraphs during the Civil War in the United States', 2 vols., (Jansen, McClurg & Co., 1882), pp. 97–98
10 BOND, W.: 'Electronic ambush of the stock market', *New Scientist*, 1976, **72**, pp. 323–325
11 VON WEIHER, S. and GOETZELER, H.: 'The Siemens Company—its historic role in the progress of electrical engineering' (Franz Steiner Verlag, Wiesbaden, 1977)
12 'Letter from Sir Lyon Playfair', *J. RSA*, 1867, **15**, p. 477
13 BITARD, A.: 'L'Exposition de Paris 1878', Libraire Illustree, *J. Hebdomadaire*, Paris, 1878
14 DAUMAS, A.: 'A history of technology and inventions, Vol. 3' (John Murray, London, 1980), pp. 384–385
15 'The Illustrated Paris Universal Exhibition', *Illustrated London News*, 1878, p. 360
16 'Paris 1878 International Exhibition', *The Electrician*, 24 August, 1878, pp. 166–167
17 'Jablochkoff Candle', *Telegraph. J. Electr. Rev.*, 1879, **7**, p. 10
18 DREDGE, J.: 'The Paris International Exhibition of 1878' (compiled from *Engineering*, London, 1878)
19 'Report of the Joint Committee of the Royal Commission for the Paris Exhibition of 1878 and the Society of Arts'. RSA ref. 069, London, 1878
20 HUDSON and LUCKHURST (Eds.): 'Report of the artisans at the Paris exhibition of 1878', *J. RSA*, 1878, 689pp.
21 NORTON, F.H.: 'Illustrated register of the Centennial Exhibition, Philadelphia 1876 and the Exposition Universelles, Paris 1878' (N.Y. American News Co., New York, 1878)
22 SANDHURST, P.T.: 'The Great Centennial Exhibition 1876' (P.W. Ziegler, Philadelphia, 1876)
23 'Turin Exhibition showing Gaulard and Gibbs system', *Telegraph. J. Electr. Rev.*, 1884, **15**, p. 364
24 'The secondary generators of Gaulard and Gibbs', *Electr. Rev.*, 1885, **5** (23), Feb 7th, p. 1
25 S 'The Turin Exhibition', *The Electrician*, 1884, **12**, p. 578
26 'First Westinghouse experimental system at Great Barrington, Mass', *Electr. World* New York, 1886, **8**, p. 271

Chapter 8
Influence of electrical power generation

8.1 Public electricity supply

The period from 1873 to 1900 represents the formative period for the electricity supply industry in Europe and America which is reflected in the extraordinary number of significant exhibitions held at this time. It began with the stabilisation of the mode of operation of the arc lamp and ended with the widespread availability of incandescent lamps and an electricity supply to drive them.

Considerable public interest was shown in electric lighting during the 1880s, the extent of which was formalised in Britain by the passing of several Acts of Parliament, particularly those of 1882 and 1888. The 1882 act created formal procedures for establishing public electricity supply undertakings. Unfortunately this Act also included a clause allowing the Local Authority to purchase compulsorily the licenced undertaking after 21 years, and for some time this acted as a disincentive to investment. The 1888 Act was designed in part to alleviate this by extending the right to 42 years, at '*a fair market price at the time of purchase*', and was followed by a rapid growth in public electricity supply schemes [1, 2].

The arc lamp, as a method of exterior public lighting, had been known since the early 1800s and was converted into a practical device by Staite and Petrie in 1846 [3]. It proved unsuitable for domestic lighting, however, but was applied with great success to lighthouses, and several examples were shown in the 1862 and later exhibitions (see also Chap. 9). Energy sources presented a problem as battery power was expensive. Experiments to drive the lamps, somewhat inefficiently, from magnetoelectric generators had been shown at the British and the French Exhibitions of 1862 and 1867 and were applied successfully only in the case of lighthouse operation [4]. Following the invention of the incandescent lamp independently by Swan and Edison in 1878, the demand for these lamps in domestic situations became general and with it the need to provide an efficient mechanical source of power in place of expensive batteries.

8.2 The Gramme dynamo

The two ideas leading to successful power generation, self-excitation and the ring armature, reached their climax with the development of Gramme's dynamo in 1871 (Figure 8.1) [5, 6]. This was the first design of machine capable of producing a continuous current of reasonable power at an acceptable cost and weight, and its use made large-scale generation of electricity possible.

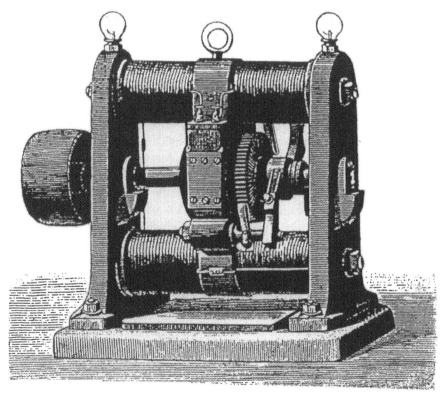

Figure 8.1   The Gramme Dynamo of 1884
(Bowers, 'History of Electric Light & Power', 1982)

Self-excitation in generator design is the use of a current produced by the generator to energise its own field windings. It had been shown by Henry Wilde in 1863 that using a separate exciting generator for this purpose could lead to much improved efficiency in electric power generation and the permanent magnet previously needed could be dispensed with [7]. After a number of attempts by several experimenters in the 1850s and 1860s, Wheatstone and Siemens independently announced a realistic solution by relying on residual magnetism remaining in the pole pieces to initiate the excitation operation and described this in papers to the Royal Society in 1867 [8, 9]. Zénobe Gramme later applied this idea of self-generation to his machine, which had its armature constructed in the form of a ring of iron with the coils wound toroidally around it (see Figure 8.2). This combination produced a stable and highly efficient current generator which had an immediate and widespread commercial success.

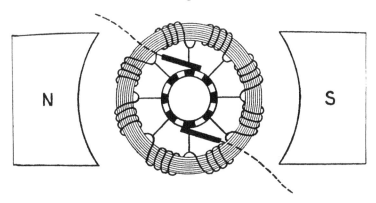

Figure 8.2   *The ring armature*
(Fleming, 'Fifty years of electricity', 1921)

The effects of public lighting were startling. In a short report made in the *RSA Journal* in 1873 the writer notes,

> '... *experiments made recently in the Westminster Road with the new electric light developed by M.Gramme of Paris ... producing illumination equivalent to 8,000 candles enabling print to be read at 270 m ... the light burnt with a steadiness which we have never seen equalled with an electromagnetic machine*' [10].

The first street in Paris to be lit by electricity was the Avenue de l'Opéra, the event made to coincide with the 1878 Exhibition (see Chap. 7).

In Europe, and particularly in Britain, interior lighting had been by gas distributed very effectively through the gas mains in towns. Towards the end of the 1870s, however, the gas companies were becoming alarmed at the possibility of electricity displacing gas as an illuminating medium. This was countered by the reassuring tone of *The Engineer* in its 9 November 1877 issue, which stated that

> '... *Electricity for domestic illumination would never, in our view, prove as handy as gas. An electric light would always require to keep it in order: a degree of skilled attention which few individuals would possess.*'

This was, of course, before the carbon filament incandescent light had appeared. Nevertheless naked gas flames were already considered by many to be offensive in terms of the odour and heat given off and additionally the use of gas was regarded as a safety hazard in large public buildings, such as theatres. The invention of the gas mantle in 1878 by Auer von Welsbach enabled gas to compete well with the feeble light produced by the early carbon-filament lamps then becoming available. But this was little more than a rearguard action, and by the 1900s the change towards electric lighting was well under way with improvements in materials used in filament lamps, culminating in the use of efficient osmium, tantalum and tungsten lamps, which almost entirely replaced gas as a lighting medium.

The immense latent demand for electric lighting released by Gramme's invention brought major changes in the content of technical exhibitions after 1873. Not the least of these was the need to establish, often substantial, generating stations on the exhibition site, and these stations themselves became important exhibits of great public interest. In was not until large public generating stations with surplus capacity became widespread in the early 1900s that these exhibition power stations ceased to be a visitor attraction, although for some time the substations on the exhibition sites, with their control equipment, were open to public view. The extraordinary use of lighting in these exhibitions made for spectacular effects, as may be seen from Table 8.1. In other areas, too, the new generators were being applied to a multiplicity of purposes: heating, measuring, communications, transport, electrochemistry and industrial use.

A satisfactory electric motor, obtained by reverse operation of a Gramme dynamo, was being widely applied in industry. A visitor at Gramme's factory at the outskirts of Paris in 1874 would have seen

Table 8.1   Lighting spectacular

Exhibition	Year	Lighting	Features
Paris	1881	200 arc lamps	First major use of
		2220 incandescent	electrical illumination
St Louis	1884	5000 incandescent	First use of knife switches
Liverpool	1886	440 arc lamps	Lighting of Brunel's
		3000 incandescent	'Great Eastern'
Edinburgh	1886	234 arc lamps	House lighting
		3200 incandescent	
Manchester	1887	512 arc lamps	High power (2000 cp (2034 cd))
		3570 incandescent	arc lamps
Paris	1889	20 000 incandescent	Interior illumination
Chicago	1893	8000 arc lamps	Use of searchlights
		130 000 incandescent	
Paris	1900	5000 incandescent	Coloured incandescent light
Buffalo	1901	200 000 incandescent	Fountain illumination
		900 arc lamps	
San Francisco	1915	370 arc searchlights	Floodlighting exterior coloured
		500 floodlights	surfaces
Barcelona	1929	4000 gas-filled	Mobile coloured lighting
		150 coloured incandescent	
		24 searchlight	
Chicago	1933	33 000 incandescent	Programmed (manual)
		100 arc searchlights	coloured lighting
		Neon & mercury vapour	Thyratron control
		tubes (30 000 m)	
Paris	1937	600 floodlights	Use of coloured metallic
		700 high power incandescent	vapour lamps for
		5000 incandescent	floodlighting
		500 mercury lights	
		750 floodlights	(for Eiffel Tower)
		8 km luminous tubes	(for Eiffel Tower)
San Francisco	1939	2300 coloured fluorescent	Dynamic control and
		8000 white floodlights	synchronised fountain
		ultraviolet & mercury	control

'... *the whole of the lathes, tools and other machines driven by connecting one of his small lighting machines by means of a belt to shafting, from which the steam engine was disconnected, which machine acting as a magnetic engine was driven at a speed of 815 rpm by a derived current from one of his larger machines which was producing on a second circuit at the same time a light of 2,400 candles*'.

We examine the effect of the new electrical devices on the content of large international exhibitions in the next two Chapters. The benefit of these new inventions for the general public was not lost, however, on the organisers of smaller international events, and the electrical content of some of these is discussed below, commencing with the widespread effect of the new electric motor on the exhibition-going public.

8.3 Electric traction at public exhibitions

An electric motor, used for traction, was first shown at the Berlin Exhibition of 1879, where a small Siemens & Halske electric locomotive hauled cars having back-to-back seating for 18 passengers around a narrow-gauge circular track 300 m long (Figure 8.3). Described by a journalist of the time as '*Eine Bahn ohne Dampf und ohne Pferde*', it created a sensation. Over 100 000 passengers were carried around the track during the period of the exhibition [11]. The railway operated from a stationary generating plant at a potential of 90 V DC with current collected from a third rail and the running rails acting as the return circuit. The train was originally designed for use in a mine tunnel and a complete railway was installed later by Siemens in 1882 in a coal mine in Saxony. A Siemens electric tram running along the Champs Elysées was demonstrated for the 1881 Paris Electrical

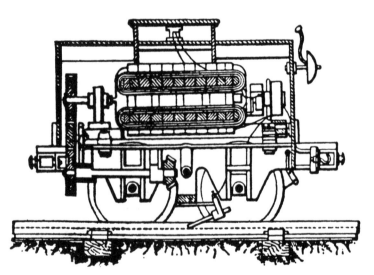

Figure 8.3   Siemens & Halske electric train shown at the Berlin Exhibition of 1879 (Engineering, 1879, 10)

Exposition, and in the following year the same company installed a more extensive tram service in Berlin between Gross Lichterfelde and the Royal Cadet College, south of the city. Each carriage had a motor under the floor connected to the wheels through a belt drive.

The first British electric railway did not open until 1883, when it provided a limited service along the sea front at Brighton.

Electric traction was sufficiently advanced in North America for it to play a significant role in the Chicago Railway Exposition of 1883. Here the Electric Railway Co., founded by Field and Edison that same year, built a 3 t 15 km/h electric locomotive which operated in the main exhibition hall and conveyed thousands of passengers over a 600 m track. This was followed the next year by an electric railway used for a similar purpose at the Toronto 1884 Exhibition in Canada. A 30 hp Van Depoele dynamo (used as a motor) pulled three cars, each with a capacity for 60 passengers, along a track, extended up to 2.5 km in 1885 in order to link with a horse-drawn tram from the city of Toronto.

A decade later at the Brussels Exposition of 1897, F.B. Behr of the Lartigue Railway Construction Company showed his monorail train which ran along a 5 km elliptical track at over 120 km/h. A similar system was later installed to run between Liverpool and Manchester in 1900. It consisted of a coach seating 64 people, straddling an overhead monorail and travelling at 56 km/h between the two cities, powered by a 650 V DC supply from Warrington Power Station [12]. At this same exhibition electric vehicles were included in its 'concours d'automobiles' together with early internal combustion motorcars [13]. By the end of the century electric vehicles were enjoying a lead in the development of personal transport, soon to be lost to the internal combustion engine. At the Automobile Club's Show in Madison Square Gardens, New York, in 1900, steam, electric and petrol cars were shown and electric vehicles far outnumbered other types in performance as well as number, as may be seen from a contemporary report:

'During the show a number of contests were arranged to compare the manipulating characteristics of the different types; and in every case the electric vehicle proved its superiority' [14].

8.4 Electric power generation at public exhibitions

It was, however, the application of electric lighting at these exhibitions towards the end of the 19th century which was to demonstrate the power of electricity to attract large crowds. For the first time it enabled the exhibition to remain open in the evening and quickly became a major source of interest not only for the spectacular decorative effects achieved but for the manner of its production. Each exhibition needed to install its own generation plant. The main attraction lay within the building housing the generating equipment, often designated as the 'hall of machinery', and no exhibition in this period, large or small, was complete without one. Such exhibitions seemed to spring up simultaneously in all the major countries of Europe, America and as far off as the Antipodes. In Britain they were linked to Queen Victoria's Jubilee, which occurred about the middle of this period. The larger exhibitions required massive generating equipment and are considered in the following Chapter. Here we look at the smaller but no less significant exhibitions, in which electric light and power generation played such a major part, considering first a number of these in the mainland of Europe, starting with the Vienna International Exhibition of 1873, coming very shortly after Gramme's invention.

8.5 The Vienna International Exhibition, 1873

The site of the Exhibition was within Vienna's immense public park at the Prater. The buildings covered an area of 16 ha—five times larger than the Champ de Mars in Paris and easily reached from the city centre. To help the public find their way about this vast exhibition site the buildings were occupied by geographically adjacent countries so that a basic knowledge of the world's geography was adequate to deduce the location of any specific country's exhibit. The exhibition consisted primarily of a number of separate palaces, of industry, machinery, agriculture and art, together with a large number of smaller buildings. A notable building, the palace of industry, contained a large nave, 900 m in length, which enabled substantial items to be exhibited. This also contained an immense rotunda designed by the English engineer, John Scott Russell, 'twice the diameter of St Peters, Rome'(Figure 8.4). In this building was staged the first major public demonstration of the use of electric power for driving machine tools. A machinery hall containing extensive generating equipment was the location for the first public exhibition of the electrical transmission of power by means of a generator and a motor. The Gramme company, who arranged this demonstration, showed two identical machines situated 500 m apart: one used as a

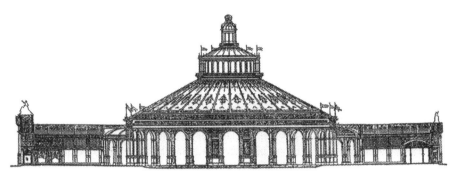

Figure 8.4   The Palace of Industry at the Vienna Exhibition of 1873
(The Electrician, 1883, 11)

generator and one as a motor driving a centrifugal pump lifting water [15]. Other generators were shown at the Exhibition, including new powerful machines designed by Friedrich von Hefner-Alteneck of Siemens & Halske. These were often used as motors and found much use in the Prussian State Mines at that time.

A few years later in 1886, the Vienna Royal Opera House became the first large theatre in the world to be lit by electricity. This was carried out by R.E. Crompton, the English engineer, using six Willans 150 hp steam engines, each directly coupled to a Crompton DC generator and capable of providing a total of 700 kW loading. The generating station was situated a mile away from the theatre, connection between the two arranged by bare copper conductors, supported on insulators, and contained in an underground culvert. The supply was used indirectly to charge 400 V storage batteries at each outlet [16].

The electrical exhibits at the Vienna exhibition in 1873 were to be found in the palace of industry. Many of these were from the Siemens Company, who contributed in the area of signalling, telegraphy and measurement. These included magnetoelectric railway signalling, Jaite's telegraph instruments, dial telegraphs, Meyer's autotelegraph, a magnetoelectric railway whistle, Bréguet's magnetoelectric exploder, various galvanometers and voltmeters and a range of Gramme generators.

8.6 Barcelona and Brussels, 1888

Two small but significant exhibitions took place at almost the same time in 1888 in Spain and Belgium. The Barcelona International

Exhibition was the first of its kind ever held in Spain, and considerable efforts were made by the Compania Espanola de Electricidad to equip the exhibition fully with lighting powered by its own generators within the grounds, although in the event three other countries also contributed to the lighting arrangements: America, Britain and Belgium.

The Spanish company already had in operation in Barcelona City a central lighting plant providing eight 2500 cp arc lamps and 500 incandescent lamps for public lighting with a smaller number for private domestic users. At the exhibition the company employed ten dynamos supplying current for 120 arc lamps and 3400 incandescent lamps located in the gardens, art gallery and the science section. The two steam engines driving these dynamos were massive, having 7 m diameter flywheels. Other generators shown at the exhibition included those of R.E. Crompton and G. Neville of Liverpool and the Ganz Company of Budapest, but the biggest installation was that of the Compagnie Continentale Edison, recently established in France. This consisted of two 200 hp engines directly coupled to Weyher– Richemond dynamos having a combined output of 800 A DC at 125 V. At the time these two machines were the largest in use in Spain.

Further generators from the Anglo–American Brush company had the specialist task of lighting the ornamental fountains in the grounds. Three high-pressure boilers provided steam for a vertical Bowett and Lindley engine from Britain, driving a pair of Victoria dynamos. An underground cable ran from the dynamos to a room built under the fountains from which the coloured lights could be controlled. Each of the arc lamps used for this purpose was of 4000 cp. From the centre of the fountain a jet of water rose some 40 m and was supported by 14 surrounding jets. The spray produced was also illuminated with coloured light. This display, which caused a sensation at the exhibition, was the forerunner of even more spectacular exhibition lighting, using water and coloured lights, to be seen at the 1929 Barcelona Exhibition (see Chap. 11).

This was at a time when considerable interest was paid to the 'consumables' of arc lamps, and a variety of carbon rods were exhibited of all sizes and types. Telephone equipment was shown from the Telegraph Works Company, the Electrotechnische Fabrik of Leipzig and the Societé Générale des Téléphone of Paris. Finally three large electric organs were established in the fine art gallery by the manufacturers Señor Amerzna of Barcelona [17].

The Brussels Science and Industry Exposition of 1888 also made much of its electricity supply and, in addition to lighting the display buildings, provided a large arc lamp display of 24 2000 cp Thomson–Houston lamps for its theatre, and a further 137 20 cp incandescent lamps in the restaurant. As noted with some of the jubilee exhibitions, the organisers were still a little hesitant about the use of dynamos and employed a backup set of secondary batteries switchable for the restaurant supply if needed [18].

At this Exposition an extensive demonstration of the Van Rysselberghe telephone system was given. This was an arrangement much favoured by the telephone companies of the time, which by the addition of an earth return were able to use the twin metallic telephone lines as one line of a telegraphic system, thus allowing both Morse transmissions and speech to be carried. It was extremely popular with railway companies. Over 2000 km of such wires were in operation in Europe at the time and, as Sir William Preece found during his visit to America in 1893, the system was also used extensively there [12].

8.7 The jubilee exhibitions, 1886–1888

The Golden Jubilee of Queen Victoria in 1886 was an occasion for holding a number of exhibitions in London, Liverpool, Manchester, Newcastle, Edinburgh and other towns in Britain during this year and the two succeeding ones. Some of these were large international exhibitions and, occurring at a time of rapid development in the generation and application of electrical power in Britain, featured large lighting demonstrations which were new to many people, certainly for those from outside London, and were responsible for the large attendance figures at all these events.

The period 1885–1898 was a time when almost all public exhibitions needed to install their own electric lighting systems; the public demanded it and the exhibitors relied on it. The electrical requirements for some of the larger public international exhibitions were immense, but all organisers managing large or small exhibitions in this period set to the task with enthusiasm and made extraordinary efforts to give the public what it wanted. This was the 'heroic age' of electrical engineering, and nowhere was it seen more impressively than during the Jubilee celebrations, where a whole gamut of exhibitions was mounted in London and provincial towns throughout Britain.

8.7.1 London, 1886

London celebrated the Jubilee with an exhibition held at the Royal Horticultural Society Gardens, South Kensington, commencing in May 1886. It did this through the 'Colonial and Indian Exhibition', the first of the 'Empire' exhibitions, discussed later in Chap. 12. It was also considered as an occasion in London to celebrate the Jubilee of the reign of Queen Victoria under the promotion of the Prince of Wales, later to become King Edward VII. It was essentially an exhibition emphasising colonial economic, cultural and political life, and as such had little place for industry other than local raw material and craft activities. The sole technical exhibit displayed was the steam power and generator installation required to supply the whole of the indoor lighting of the many halls and approaches to the Exhibition. This was suitably immense, amounting to a steam power of 1200 hp provided by nine sets of engines of various types [19].

The dynamos for the incandescent lights, mainly supplied by the newly formed Edison and Swan company, were DC machines manufactured by Mather and Platt of Manchester, who were responsible for much of the generating equipment used in the Jubilee exhibitions. Three of these operated to give an output of 105 V at 320 A, whilst a fourth was run at a higher speed to provide 130 V for charging an accumulator bank. A further two machines provided 55 V at 280 A for arc lighting. An opportunity was taken by the Edison company to carry out long-running and efficiency tests on these generators, with recorded efficiencies of 93% attained [20].

8.7.2 Liverpool, 1886

The first provincial jubilee exhibition (and also the largest), the *'International Exhibition of Navigation, Travelling, Commerce and Manufacture'*, known locally as the 'Shipperies', was opened by Queen Victoria also in May 1886. The Shipperies consisted of a series of buildings or 'courts' set in 14 ha of Waventree Park, the principal building containing a nave of ~380 m in length with a central dome of 30 m in height, which, with several crossgalleries on either side, made it an impressive building, albeit of a temporary nature (none of the buildings were permanent structures, being constructed specifically for the event so that, unlike many similar exhibitions, it left behind no permanent legacy in the form of a museum, concert hall or art gallery). One of these buildings had been used previously as the main building for the Antwerp Exhibition of 1885, and had been taken down, transported and re-erected at the site of the Liverpool Exhibition.

Three distinct features of this exhibition were: (1) an outstanding collection of maritime models, almost 200 of which were contributed by Lloyds in recognition of Liverpool's maritime history; (2) a large building containing 'machinery in motion', which demonstrated heavy milling operations, screw-cutting lathes, planing and drilling equipment and other industrial contributions; and (3) the electrical power plant, provided by the Liverpool Electric Supply Company. This provided current for 400 arc lamps and up to 3000 incandescent lamps. The compact generator installation for the exhibition was contained within a small shed 24 m square, shown in Figure 8.5. On the right-hand side of the picture will be seen three 30 hp compound engines each driving two Brush arc lamp dynamos, two 20 hp engines each driving one 50 arc lamp dynamo and one 20 hp engine working one 500 incandescent lamp machine. On the left-hand side were three 40 hp boilers, two 40 hp engines, and four dynamos supplying a total of 2000 incandescent lamps, all at 115 V DC. Four vertical engines of 4, 8, 10 and 14 hp drove four Crompton dynamos, supplying, respectively, 50, 100, 200 and 300 incandescent lamps. Some of the incandescent lamp machines were also used to run arc lamps in parallel circuits of two lamps in series [21]. In a separate exhibit by this company were several sets of plant with gas engines and accumulators shown as being 'especially suitable for private house lighting' [22].

To coincide with the exhibition Messrs Lewis and Company arranged the charter of the 'Great Eastern', Brunel's great cable-laying ship, and brought this to Birkenhead. It was still the largest ship afloat and was illuminated from the middle of May 1886 by the Jablochkoff and General Electric Companies. On the deck were 40 400 cp Jablochkoff candles mounted on standards 4 m high. A single cylinder horizontal steam engine which in the daytime drove the bilge pumps, at night was employed to drive two generators, with a third acting in reserve. The main cable tank was converted into a music hall with seating for a thousand people, and in the evening was cleared for public dances, *'by electric light'*.

8.7.3 Edinburgh, 1886

In 1886 the Edinburgh Merchants Association organised an *'International Exhibition of Industry, Science and Art'* [23]. This was held in newly built exhibition buildings in West Meadows, a substantial 7.3 ha site. The largest of these, the Grand Hall, capable of seating 9000 people, continued to be used after the exhibition.

Again a striking feature was illumination by electric light, still an

*Figure 8.5   Liverpool International Exhibition Generating Station, 1886
(Holmes, 'Practical Electric Lighting', 1887)*

attraction to the public at a time when few buildings and even fewer homes enjoyed this facility. The exhibition made use of 3200 incandescent lamps and 234 arc lamps, providing a total illumination of 725 000 cp from 14 dynamos. These were driven by nine steam engines having an aggregate power of 235 hp. The exhibition lighting was provided by Thomson–Houston and the Brush Electric Light company. This latter was to become a very important provider of lighting in exhibitions in Britain and abroad. Also driven by electricity in the grounds was an electric railway, presented by the North Metropolitan Tramway Co., running between the western and eastern entrances, a distance of about half a kilometre (fare 2d in either direction).

Within the buildings three classes of exhibits featured electrical artifacts.

These were:

Class 6:   prime movers which included electric motors as well as powerful steam engines

Class 8:   railway, tramway and vehicular appliances, particularly electrical signalling apparatus. Four steam locomotives were shown in the grounds, lent from the Scottish and British railways, which converged at Edinburgh

Class 11: scientific appliances featuring electrical illumination.

Class 6 included a 'machinery in motion' display spread over nine courts and occupying about 1200 m². The stationary steam engines used were applied to three lines of overhead shafting driving a variety of mechanical devices, including a number of dynamos for light current generation. Here a Glasgow firm of J.D. Andrews was exhibiting a '40 light' dynamo and a '150 light' dynamo, indicative of the absence of agreement on standards for electrical power—a 16 cp incandescent lamp was widely accepted as a useful unit of measurement by very many customers and referred to simply as 'a light'. The firm also exhibited arc lamps, searchlights and telegraph instruments. Similar items were shown by other Edinburgh and Glasgow firms. Several nautical electrical systems were shown, including dynamos and searchlights for torpedo boats and a combined 20 hp engine and dynamo, '... for use in steamships'. Alongside steam locomotives in class 8 were shown a number of electrical fittings including a new electromagnet contact breaker for passenger trains and, by Drew & Barnett of London, a Landau carriage 'fitted with an electric bell'. In Class 11 was shown Murrow's Rotary Loom with an electromagnetic driven shuttle. An indication of the way public interest in electric lighting was growing is shown by a pamphlet, available at the exhibition, entitled 'The electric light in our houses', priced 1d.

In the Grand Hall, despite the massive amount of power generated at the exhibition Messrs Thomas Smith of Glasgow were reported as being 'unable to obtain the necessary power to drive a dynamo' and hence 'were under the necessity to resort to the use of a battery', for their demonstration of electroplating. Electroplating, although the oldest industrial application of electricity, had been rarely demonstrated at a public exhibition, and the Edinburgh example was one of the few in this category [24].

This was the largest exhibition of its kind then staged in Scotland and the first to use electric lighting on such a grand scale. It was opened by Prince Albert and attracted 2 million visitors before it closed its doors after 6 months.

8.7.4 Manchester, 1887

The Royal Jubilee Exhibition at Manchester was opened by the Prince of Wales in February 1887. The site covered a total area of 4.8 ha, of which almost half was given over to the 'machinery in motion and general engineering' section. This consisted of one long building and a projecting addition, extending the machinery building to Manchester railway station. The building was surrounded by galleries from which the public could view the exhibition, *'without descending amongst the machinery in motion'* [25]. Its design owed much to the ideas of Paxton and the railway engineers of the period, with its wrought iron structure and plate glass roof. For lighting the building was provided with 174 arc lamps, each of 2000 cp. The exhibition also contained a number of Fine Art Galleries lit by 1 200 16 cp incandescent lamps, preferred to gas in order to avoid possible damage to the paintings. Buildings elsewhere on the site, which included extensive industrial design and chemical industries, required a further 2370 incandescent lamps and 338 arc lamps, 65 of these in the grounds supported on iron masts.

The generating equipment formed part of the exhibition of 'machinery in motion'. Steam was provided by a group of ten boilers, capable of indicating about 4000 hp. Ten steam engines were installed from several companies and drove 30 generators of different makes. The disposition of this equipment can be seen in Figure 8.6. The arrangement for arc lighting was carried out by the Anglo–American Brush Electric Light Co. Ltd, whilst the incandescent lamps were

ELECTRIC LIGHTING MACHINERY AT THE ROYAL JUBILEE EXHIBITION.

GENERAL PLAN, SHOWING ENGINES AND DYNAMOS.

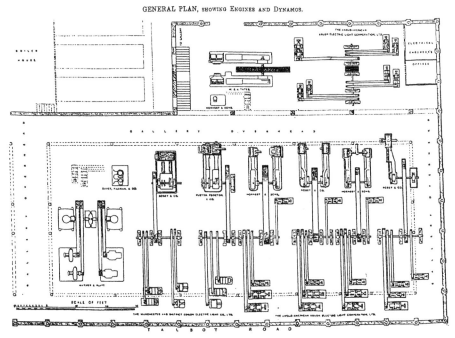

Figure 8.6 *Plan of the generating equipment at the Manchester Jubilee Exhibition of 1887*
(Engineering, 1877, 18)

installed and managed by the Manchester Edison Swan Co. Ltd, although several installations were contributed by other companies. One of the most impressive of these was by Mather & Platt of Manchester. This consisted of two vertical engines, shown in Figure 8.7, each of 200 hp and running slowly at 120 rev/min to drive directly through a step-up belt-drive a pair of Edison–Hopkinson dynamos (right-hand engine) and two Manchester dynamos (left-hand engine), the former providing a direct current at 320 A and the latter at 400 A, and all at a voltage of 100 V. A smaller machine and Manchester generator in this section conveyed current to another Mather & Platt exhibit which consisted of a large ten colour calico printing machine and an electrical singeing machine, illustrative of the important textile trade of the region.

It is interesting to note that the illumination in the grounds was not entirely trusted to electric lighting and that 10 000 'vario-coloured' oil-lamps were used, *'artistically disposed in different parts of the gardens ... (giving) a very pleasing effect to the gardens after dusk.'*

Figure 8.7   Mather and Platt generators at the Manchester Jubilee Exhibition of 1887
*(The Electrician, 1887, **19**)*

A number of static exhibits of cables, telegraph equipment, etc. were to be seen in the machinery section. The miscellany of such exhibits can be seen from the items displayed by the Electric Portable Battery and Gas Igniting Co. Ltd. of Salford:

> 'Patent dynamo-electric gas lighters, electric bells, Leclanché, Bunsen, bichromate and patent chloride of silver batteries, electric indicators, Court sets, consisting of bell, battery, wire and push, self-contained switches, one to four way, Wood, China, Brass and Pear pushes. Wire for bell and electric light purposes. Induction Coils, self-contained with sealed batteries for medical use. Show Bottles of Chemicals, as used in electric batteries. Electric cigar lighters and clock motors. Miniature Electric Bath for medical purposes. Galvanometers. Magneto-electric machines and Dynamo-electric machines for lighting. A 250 light Dynamo in a Dynamo House' [26].

The provision of telephone connection at public exhibitions was an innovative procedure at that time and was carried out with much attention to detail at the Manchester Exhibition. A 'handsome' central office and exhibit room was erected within the exhibition hall. A large map was displayed showing the cities and towns in Lancashire, Cheshire and Yorkshire with which communication was established. Sixty-one exhibitors had telephones fixed at their stands and there were a number of public telephones available in the central office, where 'three skilled female operators were kept constantly engaged'. Upwards of 100 000 messages were sent from the exhibition stands to persons in Manchester, Salford and other towns during the period of the exhibition. Telephone exhibits included, 'Taussig's patent automatic fire and burglar alarm, which was in use through being connected with the bank in the exhibition hall' [24].

The Manchester Exhibition was open for ten months and admitted $4\frac{1}{2}$ million visitors. During this time the BAAS held one of its annual meetings in Manchester and visited the Exhibition as part of its programme.

8.7.5 Newcastle, 1887

Opened by the Duke of Cambridge in May 1887, the *'Newcastle-on-Tyne Mining, Engineering and Industrial Exhibition'* was the first substantial industrial exhibition held in Newcastle. It was originated by the North of England Institute of Mining and Mechanical Engineering with many of the engineering exhibits contributed by the Sir William Armstrong Company, including ordinance, locomotives and steam boilers. The total area occupied by the buildings and gardens of the exhibition exceeded 12 ha.

The electric light machinery was contained in a separate building and consisted of 14 turboelectric generators by Clarke, Chapman Parsons and Co., with three more elsewhere in the exhibition. A total of about 390 electrical hp was generated. This was one of the first exhibitions to demonstrate the small size of generators possible, in relation to their output, due to operating them at the very high speed of 9000 rev/min, realisable with the new turboelectric generators. Over 60 arc lamps were used to light the gardens whilst several thousand 16 cp incandescent lights were applied within the exhibition buildings to give a total illumination of 100 000 cp. The supply was also used to power an electric railway in the grounds. Additionally 500 incandescent lamps were fed to a number of shops in the gardens from separate Victoria dynamos, provided by the Anglo–American

Brush Corporation, of a kind often supplied for private house installations [27].

A substantial exhibit showing the working of coal and lead mines was constructed in the grounds. This included lift shafts and winding gear with a hauling engine providing an endless wire rope for transporting the coal to the surface. The public were fitted out with electric safety lamps to tour the underground portions of this huge model.

8.7.6 Glasgow, 1888

Glasgow had already mounted an engineering exhibition in 1880, a small specialist exhibition including 'Electricity—its generation and application for lighting, telegraphy, motive power etc.' (see Chap. 6). The exhibition in 1888 was its first international exhibition designed to show engineering achievements at the birthplace of James Watt's steam engine, now a prime mover in industrial development on a worldwide basis. Despite taking place some two years after the Royal Jubilee it was described as Glasgow's Jubilee Exhibition since it provided a public opportunity for the city to display the Queen's Jubilee presents—800 of them from many lands. A full-sized model of the Jubilee Cake was also shown—it weighed over 250 kg!

The exhibition took place in Kelvingrove park, near to the University, laid out by Sir Joseph Paxton just after his work on the Crystal Palace in Hyde Park. Its buildings were arranged along two major avenues, with a machinery section at one end. This contained the electrical generators for the exhibition installed by three contractors: The Anglo–American Brush Co.; and Messrs Thomson–Houston and King–Brown of Edinburgh. The public application of electricity for lighting and transport was an exciting attraction for the people of Glasgow—it was three years afterwards before the Corporation established the first public generating station. Street lighting followed in 1893 [28]. At the intersection of the avenues was a great dome with four allegorical figures representing 'industry', 'science', 'art' and 'agriculture' at the apex of four arched dome supports and looking down on their named sections of the exhibition. There were 21 classes of exhibits including agriculture, natural resources, engineering, transportation, machinery, chemistry, etc. Under the 'philosophical instruments' section were shown on loan many of Sir William Thompson's electrical instruments, including his standard electric current balance capable of measuring 0.01–500 A, an electrodynamic voltmeter, several galvanometers, electrometers and a complete siphon recorder used for receiving signals transmitted through long submarine cables[1]. Also in this part of the Exhibition were shown a Gray–Milne seismograph recording with ink on moving paper and Gray's stationary-plate seismograph.

Further attractions were an exhibition of Glasgow's past, an illuminated electric fountain 37 m in diameter and a switchback railway. Special events were staged during the course of the Exhibition, which remained open from May to November, including a demonstration of arc-welding, the first time that this technology had been proposed for public exhibition (carbon-arc welding had been invented by Nicolas von Benardos a few years earlier in 1885). The emphasis at Glasgow was in its manufacturing capacity, and an important local industry was wire manufacture. Glasgow's first multi-block wire manufacturing machine was built in 1875. At the exhibition separate strands of 34 SWG wire were shown being produced simultaneously on a 14 die machine.

The Glasgow exhibition attracted over 5 million visitors, an attendance which, as with the 1851 and other British exhibitions, was considerably assisted by the availability of cheap railway tickets on the new railway networks.

8.8 Louisville, 1883 and San Francisco, 1894

The popularity of public lighting was exploited by the organisers of two American exhibitions in 1883 and 1894. Neither was large but both made full use of the electrical power then available to them.

The Southern Exposition at Louisville in 1883 secured limited international support, and its organisers decided quite early on to make a major feature of such electrical facilities as could be installed. It was in fact the first public exposition lighted throughout by incandescent lamps [29]. It opened for 100 days in 1883 and again for 65 days on each succeeding autumn until 1887. The capital cost of the extensive electrical installation was in this way recouped and the Exposition attracted a total of almost a million visitors over the four-year period. The electrical installation was subcontracted to the Edison Company of New York who provided 15 dynamos with

---

[1]Sir William Thomson (later Lord Kelvin) held a chair in Glasgow University for 52 years and lent his assistance to a number of exhibitions taking place in his adopted city, where he was a powerful influence in its scientific life

sufficient power to energise 4600 16 cp incandescent lamps. The power supply was also used to operate an electric railway following a narrow track circumnavigating the 18 ha grounds [30]. At one stage the track passed through an artificial tunnel lit by incandescent lamps. Most of the lamps however were used to illuminate the interior of the vast building erected for the exposition of over 63 000 m^2 and with an average height of 12 m. The lamps were the new improved high-resistance carbon filament lamps which Edison had developed in 1879 and were to be mass-produced in large quantities for his various commercial lighting installations. Edison's new lamps received all the first prizes awarded at the Exposition. It is interesting to note that Edison took the trouble to obtain an estimate for the number of gas jets (presumably the standard fish-tail burner) required to provide a similar level of illumination in the hall, and came up with a figure of 7000 to 12 000 gas jets. This was shortly before Welsbach invented his incandescent gas mantle in 1885 which would have reduced these figures to about a third. The Edison company was not so successful with its arc-lighting system, however. At another public exhibition in Cincinnati in 1885 Thomson–Houston received the top awards in arc-lighting and the Weston company challenged Edison in the electrical performance of their incandescent lamps [31].

The California Midwinter International Exposition was held in San Francisco's Golden Gate Park in 1894. Again a decision was made to make a feature of the use of electricity, but this time centred on one particular exhibit. A steel tower was constructed, a virtual copy of Eiffel's Tower from the French Exposition of 1889, but on a smaller scale—it was only 67 m high. However this tower, known as the Bonet Electric Tower, had one particular advantage over Eiffel's masterpiece in its lavish use of the newly invented incandescent lights (Figure 8.8). It has been described by Chandler as

'... *a coded metal cylinder programmed with a dazzling sequence of crosses, diamonds, rosettes and spheres across the girders of the tower*' [32].

In addition a powerful searchlight scanned the site and the surrounding countryside from the apex of the tower. During the day an electric lift brought visitors to an observation platform 64 m above the ground.

Some 3200 coloured incandescent lights were used to illuminate the tower. These were switched on and off in a sequence controlled by a mechanical drum switch, 'somewhat similar to that of a music box'. A

Figure 8.8   Bonet Electric Tower at the California Midwinter International Exposition of 1894
(Reproduced with permission from *World's Fair Mag.*, 1984, 4)

revolving cylinder 2 m long and 30 cm in diameter was made up of a number of cast-iron discs, into which were fitted protruding hardwood blocks which pressed on metal contacts as the drum revolved. These were made to operate groups of lamps in a repeating sequence every few minutes, whilst the drum was turned by a $\frac{1}{2}$ hp motor. A blower was synchronised with the motor to extinguish the arc when the contacts were broken. The combination of lights was changed several times during the day by alteration of the position of the wooden blocks.

A number of working generators were shown by General Electric, Western-Electric, Thomson–Houston and Edison. These utilised three-phase generation and transmission, of interest to the mining community in the state. Much electrical mining equipment was shown, including pumps, floodlighting and a 20 hp narrow-gauge locomotive. A demonstration was given several times per day of a 7.5 hp electric drill making holes in a block of granite for the edification of the crowd!

8.9 Australian exhibitions, 1887–1889

Australia had contributed to the international exhibition scene   fo
a number of years with several small but well attended events held
in Melbourne after 1851. The exhibitions of 1854 to 1875 (five
exhibitions) were primarily agricultural but the 1880 and 1888
exhibitions were large international exhibitions where the number
of exhibits ran into tens of thousands and attendances exceeded
1 million. The 1888 International Exhibition was a centennial event,
celebrating the discovery of Australia by Captain Cook in 1788. Several
countries took part with the United States providing the largest
contribution. Melbourne in 1888 had just established an electrical
supply industry and a decision to illuminate the exhibition buildings
and grounds by electricity was made '...*in order to avoid danger to oil
paintings contained in the galleries*' [33] (it was also a fine arts
exhibition). In their enthusiasm for the newly installed generator
capacity the local governors made the provision of electrical power
free to the exhibitors—a contributory factor to the financial deficit
realised after the exhibition closed! Some 2000 hp of generating
power was employed provided by Bush—mainly dynamos with the
capacity to supply 10 or 20 A of direct current each to a number of
groups of arc lamps[2]. A small number of compound-wound Brush
machines provided 110 V AC for about 1000 Brush incandescent
lamps each of 16 cp. The generating plant was a major exhibit within
the exhibition and was described by a jury appointed to evaluate the
exhibits as the

> '...*latest improvement in electric light ... in the adaptation of the
> alternating current system to distribute by means of transformers the
> electric current for long distances*' and '...*motive power and electric
> machinery, which have a large future before them ... demonstrated
> here for the first time.*'

This may not have been quite accurate since many of the displays and
accompanying equipment had been previously exhibited at the
Adelaide Jubilee International Exhibition of 1887, on their way to
Melbourne. The Adelaide main exhibition building was also lit by the
use of electric lights and was, at that time, the largest building in
Australia to be illuminated in this way.

---

[2]The Brush company in the United States dominated the market for the production
of arc lighting systems at home and abroad in the last decade of the 19th century due
to the extreme simplicity and reliability of the equipment

8.10 References

1 KENNEDY, M.W.: 'Electricity privatisation—engineering a culture change'. IEE divisional lecture, Newcastle, 12 May 1994
2 BOWERS, B.: 'History of electric light and power' (Peter Peregrinus, London, 1982), Chap. 10
3 WOODWARD, G.: 'Staite and Petrie: pioneers of lighting', *IEE Proc. A.*, 1989, **136** (6), pp. 290–296
4 YAPP, G.W.: 'Paris Universal Exhibition 1867—Complete Official Catalogue'. English translation, (Imperial Commission, Paris, 1867)
5 ATHERTON, W.A.: 'From compass to computer: a history of electrical engineering' (San Francisco Press Inc., 1984), Chap. 7
6 SINGER, HOLMYARD, HALL AND WILLIAMS: 'A history of technology—1850–1900, Vol. 5' (Oxford University Press, 1958), pp. 188–192
7 WILDE, H.: 'Experimental researches in magnetism and electricity', *Proc. R. Soc.*, 1866, **14**, pp. 107–111
8 WHEATSTONE, C. 'On the augmentation of the power of a magnet by the reaction thereon of currents induced by the magnet itself', *Proc. R. Soc.*, 1867, **15**, pp. 368–372
9 SIEMENS, C.W.: 'On the conversion of dynamical into electrical force without the aid of permanent magnetism', *Proc. R. Soc.*, 1867, **15**, pp. 367–368
10 'Gramme's new electric light', *J. RSA*, 9 May, 1873, **2**, p. 484
11 'Siemen's electric locomotive', *Engineering*, 1880, **10**, p. 487
12 BAKER, E.C.: 'Sir William Preece, Victorian Engineer Extraordinary' (Hutchinson, London, 1976), p. 319
13 'L'Automobilisme des Bruxelles Exposition'. The Official Organ of the Exposition, Brussels, 1897, p. 20
14 'Electric automobiles', *Electr. World New York*, 1900, **36**, p. 785
15 *Engineering*, 1879, **32**, p. 567
16 BOWERS, B.: 'R.E.B. Crompton' (Science Museum, London, 1969)
17 'The Barcelona Exhibition', *The Electrician*, **21**, 24 Aug 1888, pp. 501–502
18 'Brussels 1888 Exhibition', *The Electrician*, **21**, 10 Aug 1888, p. 449
19 'Steam power for electric lighting at the Colonial and Indian Exhibition', *The Electrician*, **16**, 6 Mar. 1886, p. 324
20 'Edison–Hopkinson dynamos at the Colinderies', *The Electrician*, 30 July 1886, **17**, p. 231
21 HOLMES, A.B.: 'Practical electric lighting', (E. & F.N. Spon, London, 1887)
22 'The Liverpool Exhibition', *Engineering*, 21 May 1886, **16**, p. 471
23 'International Exhibition of Industry, Science and Art Edinburgh 1886', Official Catalogue (Constable, HMSO Edinburgh, 1886)
24 'The Edinburgh Exhibition', *Engineering*, 4 June 1886, **16**, pp. 539–40
25 GILLES, A.G. (Ed,): 'Report of Executive Committee, Royal Jubilee Exhibition, Manchester 1887' (John Heywood, Manchester, 1887), pp. 76–77
26 'Manchester Jubilee Exhibition, Official Catalogue' (J. Heywood, Manchester, 1887), p. 158
27 'The Electric Light at the Royal Exhibition, Newcastle-upon-Tyne', *The Electrician*, 15 April 1887, **18**, pp. 510–511
28 KINCHIN, P. & J.: 'Glasgow's Great Exhibitions 1888–1988' (White Cockade Publ., Bicester, Oxon., 1951)
29 'Edison system for Louisville Exposition', *The Electrician*, 1883–84, **12**, p. 99
30 'Southern Exposition Guide 1883' (Louisville, 1883)
31 'Report on the efficiency and duration of incandescent lamps', *J. Franklin Inst.*, September 1885, 120
32 CHANDLER, A.: 'The towers of San Francisco', *World's Fair*, 1984, **4** (3), pp. 6–7
33 'Official Record of the Centennial International Exhibition, Melbourne 1888–9' (Sands & McDougall, Melbourne, 1890)

Chapter 9
Specialist electrical exhibitions

9.1 Electrical applications in industry

The generation and application of electricity as a major component in the industrial scene was now becoming significant and the time was ripe for special exhibitions or large separate sections of international exhibitions to be devoted entirely to electrical apparatus.

Some of these were directed to the more prosperous members of the general public who could afford to install lighting in their homes. Typical of these was the small Electrical Exhibition held in St Pancras Town Hall in London in 1891. Here a temporary plant of a 100 hp Marshall boiler and a Willans engine was coupled to a 48 unit (i.e. 48 16 cp incandescent lamps) Goolden dynamo providing some 500 A at 120 V for '... *a large number of lamps at the prettily decorated stalls'* at the Exhibition [1]. The Exhibition was to draw attention to the establishment of a public electricity supply in November 1891, located in the vestry of St Pancras. A key exhibit was a sample of the electrical 'mains' to be laid in the district, consisting of copper strip supported on wide porcelain blocks. New electrical meters were shown for house installation and a wide variety of electrical fittings, looking remarkably like the gas fittings they replaced with glass bead 'diffusers' and elaborate brackets. A number of public lectures were arranged, one of which was given by Sir William Preece, at that time President of the Institution of Electrical Engineers, on the comparative costs of lighting by gas and electricity.

However, the main purpose of these specialist electrical exhibitions was to demonstrate to engineers and industrialists what had been achieved and what could be achieved in harnessing this new power for manufacturing purposes, and at the same time to educate and interest the general public. This resulted in an emphasis on 'halls of machinery' and detailed measurements of the capability of the new generators which encouraged a high level of competition between manufacturers. It also released a flood of technical papers in all the technical journals on the design and application of electrical machinery of all kinds.

The last two decades of the 19th century saw very many of these specialist exhibitions held in the major cities in Europe. The International Exposition of Electricity taking place in Paris in 1881 was the largest and its success initiated a number of similar exhibitions in the following years, in London, Vienna, Munich and Prague, with provincial exhibitions held in Glasgow and Manchester, the Glasgow Exhibition taking place in 1880, a year before the Paris Exposition. This was notable for the demonstration of the new carbon filament incandescent lamps, invented by Swan in 1878. It was the first appearance of these lamps at a public exhibition, although Swan had given lectures and demonstrated his lamp earlier at the Newcastle Literary and Philosophical Society in 1879. In the United States the Franklin Society arranged their first specialised Electrical Exhibition in Philadelphia with seminal results for the electrical profession.

In the 1890s several electrical exhibitions were again held in London, Vienna and St Petersburg, with Frankfurt staging its large Electrotechnical Exhibition in 1891. In Paris in 1889 a large electrical section formed part of the Centennial Exposition, and in the United States the International Columbian Exhibition of 1893 also contained a substantial electrical engineering content.

9.2 International Exposition of Electricity, Paris, 1881

The premier specialist electrical exhibition of this period was without doubt the Exposition Internationale d'Electricité which took place at the Palais de l'Industrie in Paris in 1881 [2]. A small exhibition on the applications of electricity had taken place in Paris in 1876 but in the meantime considerable advances had taken place, fully justifying a major exhibition at this time.

The 1881 Paris Exposition was associated with an International Congress of Electricians, the first comprehensive international gathering of electrical technologists to take place. This was proposed in a memorandum from M. Cochery, the French Minister of Posts and Telegraphs, to the President of the Republic, as

'.. of great interest to clarify the state of electrical science and its applications, and to draw together and compare the results of researches' [3].

The Congress was the first body to consider the question of electrical standards on an international basis. The BAAS had already made suggestions for a universally recognised system of electrical units in 1861 and put these to the international body in 1881. They recommended the ampere, volt and ohm as practical units for current, pressure (voltage) and resistance, which were generally adopted by electrical engineers, although they did not achieve legal status until after an agreement reached at another International Electrical Congress, this time held at the Chicago Columbian World Exhibition in 1893, discussed later in this Chapter.

The Palais de l'Industrie had been used earlier as a major venue for the 1855 Paris International Exposition. The Palais contained a grand nave, 182 m in length, with a series of galleries and rooms on the upper storey. From the eastern end of this building an electrical tramcar, provided by the Siemens Company, operated to the Place de la Concorde. An overhead twin line carried the current, generated by a machine within the Palais de l'Industrie. The nave was large enough to admit full-sized signalling and other electrical devices from several railway companies. The Chemin de Fer de l'Est showed a completely equipped dynamo-metric waggon, for measuring speed, traction, brake performance etc.

Lighting and the means of generating the electric power required were, however, major features of this exposition. The impression on entering the building was described as a 'great blaze of splendour' and a 'terrific mélange of lights'. Different rooms in the Exhibition had their own set of lighting arrangements:

'a small theatre was lit by Werderman lamps, a picture gallery by a Lampe Soleil, a buffet softly lit by Swan lamps, and numerous exhibits of Ediswan lamps, each group in their own salons'.

Other filament lamps were exhibited by Lane–Fox, Maxim, Farmer and Sawyer & Man, but these never reached the commercial success of the Swan and Edison lamps [4]. Edison produced his first successful carbon filament in 1879 and patented this in England in November of that year. Swan did not seek a patent for his lamp as such, although he did patent various points of detail. It is now generally accepted that both inventors have equal claim to the design of the first successful carbon filament lamp, and the amalgamation of their two commercial undertakings in 1883 to form the Edison & Swan United Electric Light Company is evidence of this. Over the total area of the exposition there were 277 arc lamps of 116 candles each, 44 arc incandescent lamps and 1500 incandescent filament lamps—a total of 1821 electric lights, which was increased to 2500 before the end of the Exposition. At the Exposition a special commission, set up for this purpose, measured the efficiency of the various lamps (all of 16 cp) exhibited from different manufacturers in terms of generator power, with the results shown below [5]:

	Lamps per hp
Edison	12.73
Lane–Fox	10.61
Maxim	9.48
Swan	10.71

This result, together with the known longevity of Edison lamps at that time compared with other makes, was a major reason for the success of the Edison company in its various European ventures.

Just within the Palais de l'Industrie was a large lighthouse, supplied with a Fresnel lens and an electric light, worked by current derived from a De Meritens machine. It was driven by a steam engine, and although not very efficient, provided the steady current required for lighthouse work, due to its use of permanent magnets rather than electromagnets in its construction [6].

The exhibition of generating machinery was extensive, with machines shown from France, Germany, Holland, Italy, Britain and especially the United States, altogether representing a total power of 1800 hp. Despite the somewhat tardy co-operation with the British Government of the time and its wish that ' ... *such co-operation should involve no charge on Public Funds*', the British contribution was substantial. The organisation of the British exhibits was left to the Society of Telegraph Engineers to prepare at very short notice with most of the work falling on the Secretary, F.H. Webb, and the Honorary Secretary for France, John Aylmer, who lived in Paris, and later produced a splendid set of photographs of the exhibition, now in the IEE Archives, London [7]. The Brush Company showed a large exhibit, consisting of six Robey engines and a number of other machines feeding 40 arc lights. The British Electric Light Company displayed some large Gramme machines which they applied to Brockie lamps and their own incandescent lamps. Crompton and Swan also exhibited their lamps, Crompton claiming afterwards in his memoirs that ' "Cromptons" were awarded the first gold medal ever given for electric lighting plant' [8].

An outstanding generating exhibit at the Paris exhibition was provided by Edison, who had a particular interest in this field and was accompanied at the exhibition by John Hopkinson, his English consultant, who was later responsible for the construction of the Holborn Viaduct generating station in central London in 1882. John Hopkinson was Professor of electrical engineering in King's College, London, and also acted as one of the judges for this exhibition [9].

The Edison machine (affectionately known as 'Jumbo'), was first shown at the 1881 exhibition, and at that time was the largest in the world. After its sojourn in Paris it was sent to the generating station at Holborn Viaduct, London, where it became the central feature of the world's first steam-generated power station, active from 1882 to 1886, and eventually providing power to 3000 lamps in this part of London.

The generator was shipped originally from New York to Europe on the *Assyrian Monarch*, the ship that, on a previous voyage, transported 'Jumbo', a much-loved elephant from London Zoo, to appear in Phineas Barnham's 'Greatest Show on Earth' in New York, so that the further use of the name seemed quite apposite at the time [10]! The Edison machine weighed 20 t and was driven by a 125 hp steam engine to produce 900 A of current at 110 V, sufficient power to drive 1200 16 cp incandescent lamps. Later two similar installations were added to the power station. The choice of 16 cp for Edison's lamps is considered to be in accordance with his desire to market a lamp providing approximately the illumination of a gas fishtail burner, the common form of gas installation at the time. However, with the improvement in gas illumination obtained with the gas mantle of Auer von Welsbach in 1878, a range of lamps were later produced by the Edison & Swan Company having standard candle power ratings of from 1 to 1000 cp [11].

At about the same time as the opening of the Holborn power station in London, Edison was busy installing a similar generator for operation in a generating station situated in Pearl Street, New York, and serving an area within the city. Up to this time in the United States (and indeed throughout Europe), wherever electrical power was installed the purchaser operated his own plant. The opening of the Pearl Street Central Generating Station in 1882 was claimed as '...*the beginning of the Electrical Age for the United States*' [12].

Several exhibitors at the Paris exposition applied electric motors to domestic and industrial equipment. Marcel Deprêz demonstrated five motor-driven sewing machines, lathes, a drilling machine and a printing press. John Hopkinson showed his 50 kg industrial electric hoist [13] and an electric lift was exhibited by Siemens. This latter had previously been shown for the first time at an exhibition in Mannheim–Pflazgan in 1880. The Exposition also provided the opportunity to display France's first electric tramway, incorporating motors manufactured in the new Siemens–Frères works in Paris. The Siemens company demonstrated their new electric furnace, patented in 1879 to melt steel ingots placed in contact with the arc furnace formed between two carbon rods of large cross-section [14].

Apart from the lighting exhibits at the Paris exposition, important sections were given to telegraphy, in which the British exhibits were most notable. Several forms of Wheatstone's automatic receiver were shown, and a comprehensive collection of historical instruments provided by the Royal Institution and specimens of the experimental cable laid at the time of the 1851 Exhibition between Dover and Calais. A demonstration was given by the Eastern Telegraph Company of William Thomson's siphon recorder working through an artificial cable of 1900 km. The first demonstration of the siphon recorder took place in June 1870 at the London residence of Sir John Penderon on the completion of the British–Indian submarine cable [15].

Amongst the electrical measuring instruments of Elliot, Clark and Muirhead was shown an ingenious fire alarm system by Edward Bright, a premier experimenter of the time. This demonstrated eight street posts electrically connected in series, similar to those located by the Fire Brigade of London for the use of the police or public in case of fire. On pulling out a handle a resistance is included in the circuit, the value of which differs with each post. The resistive chain becomes part of a bridge circuit and the resulting imbalance is detected at the central fire station, causing a bell to be rung. Subsequent rebalancing of the bridge allowed the value of the inserted resistance to be determined and hence the location of the fire post.

In the United States section a significant number of military electrical exhibits were shown. Major Heap has left an outstanding description of this equipment in his report [16]. He describes the use of dynamos and magnetoelectric machines for military telegraphs, portable power supplies, mine exploders and torpedo apparatus.

The most significant exhibition in this section was Graham Bell's recent invention of the telephone. A special room was set aside where the visitor could pick up a pair of earphones and 'listen to' a performance, 'live', as we would now say, at the Opera House a kilometre away. Bell incorporated in his display a rudimentary telephone exchange, probably the first ever shown. Another new use for electrical transmission, demonstrated by Van Rysselberghe at the Exhibition, was his 'telemeteorograph', which telegraphed every 10 min the state of the weather at Brussels, to Paris, where it was recorded. The crude transmission of pictures by telegraph wire, shown by Shelford Bidwell, taken together with the invention of the telephone, led *The Electrician* of 1881, normally an extremely staid publication, to surmise that,

> '... *an absent lover will be able to whisper sweet nothings in the ear of his betrothed, and watch the bewitching expression of her face the while, though leagues of land and sea divide their sympathetic persons...*' [17].

The Paris Electrical Exposition was outstanding in one further respect—that of education. Unlike in some of the earlier French expositions, a deliberate attempt was made to give more explanation of the exhibits on show and particularly the historical displays provided in the French, British and German sections—a process repeated in many subsequent international exhibitions.

9.3 The Crystal Palace and Westminster Aquarium Electrical Exhibitions, 1881–1883

Later, in the same year as the Paris exposition, an International Electrical Exhibition took place at the Crystal Palace London, now located in its new home in Sydenham Hill. Although it was called 'international', most of the exhibitors were British. Fewer exhibits were shown than in the Paris Exhibition (300 exhibitors compared with 1700 at Paris), and indeed many of these were simply transferred from Paris to provide Londoners with an opportunity of viewing them. Whilst it was on a smaller scale than the Paris Exhibition, contemporary commentators seemed to agree that it was better arranged and enabled the different manufacturers' electric lights to be compared more easily [18]. The whole Exhibition was lit by electric lights supplied by numerous manufacturers with Edison lamps in prominent display. Sir Ambrose Fleming had recently been appointed as Electrician to the Edison company, and spent much of his time at the Exhibition explaining details of the Edison generating system to the public. This comprised 12 Z-type dynamos providing current for the companies' exhibits. Shortly after the Exhibition Fleming was responsible for installing one of these aboard the Admiralty's Indian troop-ship, *HMS Mooltan* [19].

The elaboration found in Victorian articles had not completely disappeared by 1881, and one of the sights of this exhibition was a chandelier shown by the Ediswan Electric Light Company. This consisted of 350 flowers represented in hammered brass and each containing at its centre an incandescent lamp. The whole structure was almost 5 m high and 3 m wide and put forward as an example which lends itself, '... *as well as, if not better than, gas to elaborate highly aesthetical designs in central and other lustres*' [20].

Another small electrical exhibition in London was the Westminster Aquarium Electrical Exhibition held in 1883. This was significant in demonstrating for the first time the AC transmission scheme of Lucien Gaulard and J.D. Gibbs of Paris, which represented an improvement in the economy of supplying a number of small loads from one common generator in addition to enabling long distance electrical transmission at high voltage and low current. In this system the primaries of all supply transformers were connected in series and the individual loads fed from the secondaries. A contemporary account states,

> '... *It has a two-fold object, first it aims at rendering it practicable for undertakers to supply current at the most economical potential permitted by the terms of their provisional orders; and, secondly, it is intended to make the user independent of the producer, and enable him to apply the current he receives to any purpose he may please, such as arc lighting, incandescence lighting, the generation of power or heat. ... If the secondary coils be made in several parts, each with independent terminal, these parts may be combined either in parallel, compound parallel or in series, according to the conditions under which the second or locally generated current, is to be employed*' [21].

It was not quite as successful as the inventors made out. Their method of connecting all the primary circuits in series did cause the loading on one secondary to affect others. Zipernowsky showed an improved system with parallel-connected primaries at the Inventions Exhibition in London in 1885, which overcame the problem (see Chap. 6).

A second feature of this exhibition was the large number of different arc lighting systems shown—some 15 in all. This reflects the difficulty at the time in arriving at a fully satisfactory system to control the adjustment of the arc gap as the carbon burns away. Professor A.B. Holmes in his book, '*Practical Electric Light*', published in the same year as the exhibition [22], describes 16 alternative systems together with eight electric candle arc lamps, most of which were displayed in one or other of the electrical exhibitions held in the last two decades of the 19th century. In addition to the ubiquitous Edison display of lamps, present at all exhibitions of the period, the new firm of Ferranti, Thomson & Ince lit the dining annexe of the Aquarium with 350 incandescent lamps of their own manufacture and the Metropolitan Brush company installed electric lighting at the Imperial Theatre in Westminster to coincide with the opening of the Exhibition.

9.4 Munich International Electrical Exhibition, 1882

Germany held the first of its two specialised International Electrical Exhibitions in Munich in 1882. We are fortunate in having a first-hand account of this exhibition by a leading contemporary electrical engineer, Sir William Preece, Engineer in Chief at the Post Office, writing in the *Journal of the Society of Telegraph Engineers*, from which much of the following account is taken [23].

The Exhibition was even smaller than the Crystal Palace Electrical Exhibition, containing only 170 exhibitors, and most of these from Germanic countries, although France and Britain made useful contributions. The items were however selected with care by the exhibitors to provide a useful view of electrical engineering accomplishments and practice at that time in Europe. It was held in a glass palace, modelled on the Crystal Palace, but smaller and having only one transept. It contained an early demonstration of a fountain illuminated by electricity and a small theatre lit by electric light.

Preece notes with approval a number of hand-driven dynamos used to illustrate experiments and capable of providing a current of an ampere or so, sufficient to light two or three incandescent lamps:

> 'These are cheap and would be a valuable asset', he claims, 'for the laboratory experimenter or lecturer'.

The large generators and dynamos on display included machines by Ruston & Proctor of England, Siemens, Brush, Neumayor and Schuckert of Nuremburg. This last organisation had the largest display of dynamos shown at the Exhibition. Shuckert sited one of these on the Iser river, some 5 km from the Exhibition, where it consisted of a water-driven turbine of 12 hp driving a DC generator to produce a current of 8 A which was used to light 11 arc lamps at the Exhibition. Another example of power transmission was given by Marcel Deprêz of Paris. A gas engine was situated at Miesbach, 55 km from Munich, driving two Gramme machines and providing DC power at 1500 V along the iron wires of a telegraph line to the Exhibition. The overall efficiency of this transmission system was considered to be about 35% [24].

Accumulators were still of interest to users, especially at small private installations, and Planté brought his display of batteries from the Paris 1881 Exposition for display. Shulze of Strasburg showed a new design of lead–acid battery in which the lead plates were covered with plumbic sulphide, which he claimed increased its efficiency considerably. A wide range of lamps were seen. Crompton showed his arc lamps, and incandescent lamps were displayed by Edison and Swan (the companies were to merge a year afterwards in 1883). Other lamps were shown by Gremer, Fredericks, Müller, and Cruto of Italy. Telegraph equipment was shown with some emphasis on railway signalling, displaying the equipment used for the Bavarian Railways and, from the Paris exposition, the exhibit of the Chemin de Fer du Nord. A surprising omission from earlier exhibitions was the lightning conductor, here seen in profusion since, as Preece points out, protection from lightning is a common feature seen in the German large buildings, whereas it is seldom seen in Britain.

A major feature which was new to specialist exhibitions of this kind was the arrangement made for professional testing and recording of the electrical performance of many of the exhibits, foreshadowing the more elaborate arrangements to be seen at the Frankfurt Electro-technical Exhibition of 1891. A long gallery was set aside for this purpose containing precision resistors, thermometers, galvanometers, voltmeters, photometers, an electric dynamometer and a torsion galvanometer. Calibration of these instruments was effected by fundamental weight measurement of the deposition of zinc during an electrolytic process. A group of professors from universities and technische Hochschule throughout Germany took part in this process, and the results were recorded by Dr Beetz, the head of the Exhibition. The exercise is interesting for the light it sheds on measurement units in use throughout Europe at that time. Whilst current was measured in amperes, and electromotive force in volts, resistance was recorded in Siemens units, rather than ohms, although a conversion to the latter is a simple arithmetic ratio. Measurements involving horse power, such as candle-power per horse-power, a common requirement for lighting installations at that time, were made in the French cheval vapeur (equivalent to 736 W, compared with the 746 W of the British horse-power). To add to the complexity the definition of English candle power was used for the measurement of incandescent lamps, whereas the French bec carcel[1] was used for measurements on arc lamps.

9.5 Vienna International Electrical Exhibition, 1883

This was held in Vienna's public park, the Prater, the venue for the 1873 Exhibition, again using the Palace of Industry . An electric tram ran through the park [25]. Sixteen countries took part in the Exhibition. Much use was made of the central feature of the building, the rotunda (see Figure 8.4), with its capacity of almost half a million cubic metres. This was to be used for a number of public events until destroyed by fire in 1937 (the same fate that befell the Crystal Palace at Sydenham one year earlier). The three galleries of the rotunda, located at heights of 24, 48, and 66 m, were filled with exhibits and the size of the rotunda permitted the installation of interior lighting comprising 20 Sun arc lamps (lampe soleil), each of 1000 cp and first seen at the Paris 1881 Exposition. These lamps occupied a position midway between an arc and an incandescent lamp. A large radiating surface is obtained by raising a piece of refractory material, generally marble, to incandescence by means of an electric arc. This is shown in Figure 9.1. The marble is drilled on two sides to admit the carbon rods, one of which has a small hole along its centre to allow a thin carbon rod to be pushed through and make contact with the other carbon in order to initiate the arc, after which it is withdrawn. The diagram also shows the elaborate nature of the exterior lamp design with its protective glass globe. This enclosure not only softened the light, making it more comparable in quality to incandescent light, but also increased the life of the carbons—an important factor [26]. The marble lasts for only some tens of hours and like the carbon rods needed frequent replacement.

The British Commissioner had some difficulty in obtaining sufficient British entries for this exhibition and, as for the Paris 1881 Exposition, appealed to the Society of Telegraph Engineers for further exhibits. Several members responded and as a result a splendid collection of early telegraph apparatus was assembled, including samples of submarine cables, an original Wheatstone Bridge, and measuring items from Sir William Thomson's collection.

---

[1]Equivalent to 9.6 international candles, whilst the international candle is equal to 0.98 of the old English standard candle

Figure 9.1   Sun arc lamps (lampe soleil) at the Vienna International Electrical Exhibition, 1883
(*The Electrician*, 1883 **10**)

Few new exhibits were to be seen however. One was a powerful electric arc lamp by M. Krizik, the inventor of the Pilsen arc lamp. In this new version a single lamp was claimed to produce a light equal to 20 000 candles. It was driven by a dynamo linked to a 60 hp engine. Improved versions of exhibits already seen at Paris, Crystal Palace and Munich formed the major part of this Exhibition with many of the features of the Exhibition already familiar to the exhibition-going public. M. Deprêz from France again showed the transmission of electric power over a distance of 56 km using a water-powered dynamo. As with his Munich demonstration the previous year he transmitted the power over iron telegraph wires and thereby achieved a low efficiency. An electric railway was arranged to run through the principal avenue of the Prater to the Exhibition. Several Viennese theatres were connected to telephones in the rotunda to enable their performances to be heard by the visitors. The highly successful Hughes printing telegraph, now widely used throughout the Continent, was shown, and the Edison company provided its usual large variety of exhibits, the only new example being the recently patented electric cigar lighter, soon to be a fixture in all large American automobiles. The short developmental time between electrical exhibitions in this period was, of course, the reason for this lack of new exhibits. In an editorial to *The Electrician* in 1891 the Editor commented, apropos the difficulty in finding something new to say about these frequent exhibitions '*... may we never have another Electrical Exhibition in this century!*' [27]. Three more were arranged, however: the Philadelphia International Electrical Exhibition; the Frankfurt Electrotechnical Exhibition; and the second Crystal Palace Exhibition, about which the editors of *The Electrician* and other technical journals were later to enthuse.

9.6 Philadelphia International Electrical Exhibition, 1884

In 1884 a small International Electrical Exhibition was presented in Philadelphia through the Franklin Institute which, as we noted earlier, had been arranging its own private exhibitions for a number of years. The Philadelphia Exhibition is recognised as a landmark signalling the beginnings of professional engineering for the electrical industry in America that came with the formation of the American Institution of Electrical Engineers (AIEE) in 1884 [28].

A new exhibition building was erected and used subsequently for later events. This was quite large with a domed glass-covered central hall 31 m wide by 62 m long, with two smaller arches on either side, flanked by 19 m gothic towers at each of the four corners of the building [29]. Over 2000 exhibits were shown, provided by 216 exhibitors from the United States and abroad, and the exhibition was open for just six weeks. Seven sections of exhibits were defined covering every aspect of electrical science, including a special section 5 on terrestrial physics. In the 26 classes contained in the two applications sections (for low-power and high-power apparatus) separate classes were allocated to such detailed uses as 'applications of electricity to dentistry' and 'electrical toys' as well as the more usual applications such as telegraphy, generators and the electric motor [30].

The Brush DC generators shown at this Exhibition incorporated a greatly improved design of armature in which the cast-iron ring generally used was replaced by a core built up from laminated steel plate. The improvement in efficiency obtained enabled the load to be increased from 40 to 65 lights in one design and led to Brush installations being widely used in the United States for street arc lighting [31]. A few years earlier in New York City, 22 Brush arc lights were installed in Broadway, displacing 500 gas lamps, with the price paid by the city being 20% less [25]. A range of generating equipment was also shown by the Weston Company, which had earlier specialised in constant potential generators required by the electroplating industry, particularly in England [32]. Here was exhibited a similar design of generator for arc and incandescent lamp installations.

The success of the Edison carbon filament lamp encouraged many other manufacturers to produce their own designs. One of the most successful of these was the 'zigzag' carbon filament lamp designed by Dr Edward Wilson, who later entered the electrical instrument field. These lamps were exhibited for the first time at the Philadelphia Exhibition in 1884, their principal advantages being compactness and improved methods of manufacture.

Although little notice was taken of it at the time, the Exhibition was notable for the first public display of a special incandescent lamp showing the 'Edison effect"—the flow of electrons across a vacuum tube, which preceded its application by J.A. Fleming in 1904 as the diode valve, and the beginning of modern electronics.

This exhibition also saw the début of Frank Sprague's industrial electric motor, which was to find wide use in the following decade. It appeared at a time when the Edison Company in New York was endeavouring to attract more daytime loading for its electric light generating stations. The Sprague motor was recommended by Edison in the following glowing terms:

'The Sprague motor is believed to meet ... all the exigencies of the case (for day-time loading) and the Edison Electric Light Company feels that it can be safely recommended to its licensees as the only practical and economic motor existing today' [25].

One consequence of this exhibition was the interest it engendered in the United States Electrical Commission, established by Act of Congress. This body issued invitations to the large number of scientists—American and foreign—expected to visit the Exhibition, to attend a National Conference of Electricians in Philadelphia, which led to the foundation of the American Institution of Electrical Engineers, predecessor of the Institute of Electrical and Electronics Engineers (IEEE) [33].

A further activity, arranged in conjunction with attendees of the conference, was the most complete and extensive independent series of tests of electrical apparatus and appliances that had yet been undertaken in the United States. The results of these measurements were recorded in 29 separate reports and were later published in the *Journal of the Franklin Institute* over the succeeding 12 months [34]. The extent of these measurements provides a useful contemporary record of the state of the electrical industry at the time, and may be compared with similar work carried out in Munich and Frankfurt, described earlier in this Chapter, and Guilbert's record of electrical measurements taken at the Paris Exposition of 1900 [35].

The Exhibition offered some electrical diversions, including that of the 'Electric Girl Lighting Company', who offered to supply 'illuminated girls' for indoor occasions. Young women were hired to perform the duties of hostesses and serving girls decked out in filament lamps attached to their clothing. These were advertised to prospective customers as, *'girls of fifty candle power each, in quantities to suit householders'* [36].

9.7 Frankfurt Electrotechnical Exhibition, 1891

The role of the specialist electrical exhibition, ushered in with the French Electrical Exposition of 1881, was fully established with the German Electrotechnical Exhibition, held in Frankfurt in 1891. This is not altogether surprising. During the last decade of the 19th century the German Electrotechnical Industry was growing rapidly, and by the first decade of the 20th century was the biggest in Europe, twice the size of the comparable British industry and only slightly below that of the United States [37].

The site of the Exhibition, opposite the railway station in Frankfurt, was ~82 ha in extent and included a small lake. Four different kinds of electric railway linked the Exhibition to the Opera House, Schiller Platz in the centre of the town and the river Main, which latter provided excursions on electric launches. The line to the Opera House was 1.3 km long and fairly level. A Siemens & Halske tram (Figure 9.2), already in use commercially in the Lichterfelde district of Berlin, carried 30 people in each carriage. The current was

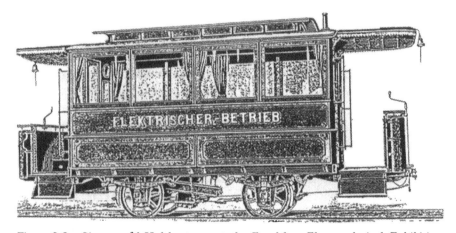

Figure 9.2   *Siemens & Halske tram at the Frankfurt Electrotechnical Exhibition, 1891*
('Allgemeiner Bericht über die internationale Elektrotechnische Austellung in Frankfurt am Main, 1891')

provided on an overhead line from the machinery hall within the Exhibition. In addition, a larger car holding 40 passengers was driven by two 10 hp motors powered from a set of 162 cells carried beneath the vehicle [38].

Details of the Exhibition are given in the two-volume official report, including an extensive report by the juries (Prüfungs– Kommission), led by Professor von Helmholtz [39]. In volume 1 the nine classes of exhibits, covering dynamos, accumulators, lighting, motors, telegraphy, telephone & signalling, electrochemical, measuring instruments and electromedicine, are described. Volume 2 gives the juries' reports. These latter are not confined to a description of the exhibits and prizes gained but set out a full-scale range of measurements on each exhibit—for example, loading factor curves, regulation, transmission losses, conversion efficiency—the work being carried out under the supervision of the leading engineers of the day from universities and technische Hochschule throughout Germany and abroad. A list of names and qualifications of the 100 professional members making up this jury is given, which include Ayrton, Fleming, Hopkinson, Preece, Kapp, Swan, Crompton and Silvanus Thomson—all from Britain. In addition to their jury work a number of the engineers and scientists gave lectures at an International Electrotechnical Congress which took place at the Exhibition (the Proceedings were later published in the *Elektrotechnische Zeitschrift*).

The heart of the Exhibition was the big machinery hall, measuring 140×60 m, part of which is seen in Figure 9.3. Here were housed 69 motors and 62 dynamos, converting a force > 4000 hp and varying in size from 1 to 600 hp. Some of the driving steam engines were massive structures; others were small gas motors (but still with 2 m flywheels) and water motors. The dynamos exhibited, both AC and DC, were from manufacturers in Germany, Switzerland, the United States, France and Britain. The machines were grouped together in the hall for each manufacturer. For example, the machinery from the Berliner Maschinen Aktiengesellschaft consisted of seven machines directly coupled to steam engines of the same manufacture as follows:

1. compound machine of 300 A, 120 V, 260 rev/min and having 6 poles
2. compound machine for a ship 140 A, 110 V, 300 rev/min and 4 poles
3. compound machine for a ship 45 A, 110 V, 450 rev/min
4. compound machine of 500 A, 120 V, 175 rev/min
5. shunt machine 100 A, 110 V, 650 rev/min and 4 poles
6. shunt machine 50 A, 110 V DC, 750 rev/min
7. shunt machine 10 A, 150 V DC, 900 rev/min used to charge a battery bank for a demonstration of house lighting.

Other exhibitors of large machines were the Deutschen Elektrizitäts–Werke of Aachen, Thomson-Housten International Electric Co. of Boston and Hamburg, Woodhouse & Rawson of London and Siemens & Halske. This last showed the largest working AC generator at the Exhibition, providing an output of 330 kW at 200 V when running at 100 rev/min and having a massive armature of 4.6 m diameter. Although the discovery of the induction motor is attributed to Nikola Tesla in the United States and the first examples made by the Westinghouse Company in that country, the first consistently success-ful commercial induction motors were those made by C.F.L. Brown at the Oerlikan works in Switzerland and those designed by D. von Dobrowolsky the chief engineer of Elektrizitäts Aktiengesellschaft in

Berlin [40]. Examples of both machines were shown at Frankfurt. A number of portable power supplies were shown. These were used for providing powerful lighting at building sites, working on the railways

Figure 9.3   The machine hall at the Frankfurt Electrotechnical Exhibition, 1891 ('Allgemeiner Bericht über die internationale Elektrotechnische Austellung in Frankfurt am Main, 1891')

Figure 9.4   Portable generator at the Frankfurt Electrotechnical Exhibition, 1891
('Allgemeiner Bericht über die internationale Elektrotechnische Austellung in Frankfurt am Main, 1891')

at night and for emergency use. The one illustrated in Figure 9.4 was exhibited by the Daimler Motor Company and consisted of a 5 hp petrol engine, a dynamo providing 65 V at 40 A, a battery of lamps, and searchlights with their control equipment, cooling water and petrol supplies.

A new feature for an electrical exhibition, carried out at Frankfurt by Oskar von Miller, was a detailed study of power transmission set up on a 3-phase line from a generator at Lauffen, on the upper Neckar river, to the Exhibition at Frankfurt. This was 170 km in length with transformers and switchgear at both ends and powered by a 356 hp turbine driving a Swiss Oerlikon generator. This transmitted 225 kW of 3-phase AC down the line at a voltage up to 30 000 V. A full technical report on the performance and operating capability for this transmission system was included in the jury's report. *The Electrician* in 1892 reported that the jury, following the completion of their tests, intended to, '... *gradually push the voltage higher and higher until the insulation breaks down*' and, at the time of reporting, had more than doubled the working voltage of the electricity supply.

It was this demonstration at Frankfurt of the ability to transmit AC power long distances at a high potential, using comparatively thin conductors, that helped to convince Europeans of the value of AC over DC for public electricity generation and transmission. At the beginning of the 1890s several AC generating stations had been constructed, including the massive power station at Deptford, south of London, and thereafter AC generation leading to a national system of electricity supply became the norm.

Although power generation formed a major theme at the Frankfurt exhibition, electric lighting was also well represented in all its forms, including new and powerful searchlights, all powered from the machinery hall. Electric motors were shown performing many tasks, for machine tools, pumping, electric fans, etc. Some 22 workshops of different trades were set up in the grounds of the Exhibition, primarily to demonstrate the use of individual electric motors to drive drills, lathes and other power tools in place of the forest of overhead driven belts usually encountered in a 19th century workshop.

The section on telegraphy, telephony and signalling was well developed, with many of the American exhibits seen previously at the Paris Exhibition of 1889. A complete telegraph station of the Königlich Bayerischen Posten und Telegraphen was created within the Exhibition. This displayed a number of perforated paper Morse transmission sets, telephones and operator switchboards. A variety of elegant domestic telephones were exhibited by several manufacturers (Figure 9.5), all from Germany, including C.E. Fine of Stuttgart, Gould Co. of Berlin and G. Wehr also of Berlin.

A spectacular exhibit was a *'captive balloon capable of seating ten persons and controlled by Captain Rodeck'*. The pulley which controlled ascent and descent was operated by an electric motor, a telephone line maintained communication with the ground and *'experiments made with a view to the steering of the balloon by electricity, and in the filling of the balloon with electrically prepared water-gas (hydrogen)'*.

9.8 The Second Crystal Palace Electrical Exhibition, 1892

Known as the 'Electrical Appliances Exhibition', the second Electrical Exhibition took place in 1892 shortly after the Frankfurt Exhibition. At the Crystal Palace mains wiring was installed by the Fowler–Waring Co. to the new Sydenham Power Station which had just been opened. The cables were laid in porcelain pipes of 13 cm bore, each pipe containing three cables carrying a 1000 V DC supply. A number of self-contained generator sets were, however, shown working in the

Figure 9.5   Domestic telephones at the Frankfurt Electrotechnical Exhibition, 1891
('Allgemeiner Bericht über die internationale Elektrotechnische
Austellung in Frankfurt am Main, 1891')

exhibition hall, the most popular for private installation being a combination of a Brush engine coupled direct to a Victoria dynamo. Some alternators by Mordey and Crompton were also in use. A very large steam dynamo provided an impressive working exhibit with its 2 m flywheel and hiss of escaping steam. This was a joint exhibit by Johnson & Phillip and Davey Paxman and Company, and featured a Kapp dynamo rated at 143 kW. The direct current generated, of up to 550 A at 205 V, was used for lighting a picture gallery by incandescent lamps, for a number of Gülcher arc lamps and for driving some of the machinery in the exhibition hall [41]. Augmenting this display were several of the outstanding exhibits transferred from the Frankfurt Electrotechnical Exhibition.

The Crystal Palace Exhibition marked the beginnings of an electrical appliances market in Britain. For the first time in Britain a demonstration of the electric iron was given by J.J. Dowsing (it had been invented earlier in 1882 by Henry Seely of New York). R.E. Crompton, one of the pioneers of the electric supply industry, exhibited early electric cookers and heaters, some of which were to reach the retail market in 1894 and resulted in a marked increase in the marketing of cookers in that year [42]. The heaters made use of the Lane Fox system of embedding bar iron heating wires in a layer of enamel on cast-iron studded plates about 30 cm square. The panels were set in highly ornamental wrought-iron screens for domestic use (Figure 9.6). To popularise the advantages of 'electric cooking' an 'all electric' banquet was arranged by the City of London Lights Co., to which many exhibition-goers were invited [43]. One author at the time even claimed that cakes cooked in this way had a certain undefinable 'electrical flavour' [44].

'Listening in' to concerts from outside the exhibition was still an item of lively interest with the public, and the National Telephone Co. arranged this on their Exhibition stand where, 'on payment of a small admission charge' the visitor could listen to music performed at theatres in London, Birmingham, Manchester and Liverpool, but omits to inform us whether this could all be accomplished on the same evening [45].

This second Crystal Palace Exhibition proved to be the last entirely electrical exhibition to be arranged in Europe, apart from a small exhibition of electrical appliances held in Como, Italy, in 1899. Most of the large international exhibitions that followed in the next few decades had their electrical sections, sometimes quite large ones, but gradually electrical artifacts became absorbed into the larger exhibition themes, as we shall see later.

Figure 9.6   Crompton's original electric fires for domestic use shown at the 1891
Crystal Palace Electrical Exhibition
(Crompton domestic appliances catalogue, 1891)

9.9 Paris Centennial Exposition, 1889

The huge Paris Centennial Exposition of 1889 included a substantial electrical exhibition [46]. The Exposition was intended to be something rather special since it celebrated the centenary of the French Revolution of 1789 and had the full support of the government of the time. Gustav Eiffel was commissioned to design and build his tower (300 m high, using 7300 tonnes of wrought iron) on the site of the Exposition, the Champ de Mars. This magnificent achievement by France's premier engineer was to stimulate the organisers of other international exhibitions to include a similar symbolic central structure in almost all subsequent international exhibitions.

The Exposition covered an area of 70 ha (40% bigger than the 1878 Paris exposition). To get around this huge site the organisers staged a light railway within an exhibition site. This was the Deceauville Railway, which proved a great success, carrying some 6 million visitors during the 5 months the Exposition was open.

On account of its form of celebration the Centennial Exposition was boycotted by Germany, Austria and Sweden,[2] but otherwise attracted a large number of foreign exhibitors taking 4400 exhibition places. Nine classes of exhibits were defined, and one of these, the 6th, was designated 'mechanical inventions and electricity'. This section contained many of the electrical exhibits familiar from earlier exhibitions. New features included illuminated fountains with coloured lights changing in accordance with a programmed relay sequence and, for the first time, an electric propelled balloon submitted by the French School of Military Ballooning. It contained a specially designed 'lightweight' battery of 435 kg and an 8.5 hp motor.

By far the most spectacular electrical exhibit was a huge display (837 m^2), within the American exhibition area, devoted to the works of Edison. Here were displayed all his 493 inventions that had earned him the title 'king of light' in France at that time. The Société Edison was by 1889 firmly established in Paris and a few years earlier had fitted the Opera House with 300 Edison incandescent lamps driven from a 50 hp gas engine in the basement. Edison's lighting display showed, amongst other exhibits, a group of 20 000 incandescent lamps arranged in the shape of a huge single lamp. In the physics area an electric pen and duplicating press was demonstrated. This produced a stencil pierced electrically with several thousand hole positions to make up a picture, through which ink was forced to effect the mechanical printing. To produce the stencil a tiny electric motor, running at about 4000 rev/min, vibrated a pin that punched holes in waxed paper at a high rate. Edison referred to this arrangement as his 'electric pen', which dates from 1876, making it one of the earliest applications of an electric motor. This section also contained a megaphone (essentially a speaking trumpet augmented by two huge resonators), a magnetic bridge which measured the magnetic properties of specimens, a weighing voltmeter, and a public

---

[2]It was billed as '...summing up all the progress that free labour dating from 1789, which is an economical as well as a political date, has accomplished during a century now drawing to a close. It is to the study of this economical situation that all nations are invited'

demonstration of his phonograph system (Figure 9.7), seen earlier at the 1878 Paris Exposition. An electric power department contained two 480 lamp machines and a dynamo with the capacity to drive a further 2500, 16 cp lamps. The lavish use of electric light within the

Figure 9.7   Edison's phonograph system shown at the 1889 Paris Exposition
(Official Guide, 'L'Exposition de Paris, 1889')

grounds made it possible to open the Exposition at night, the first major exhibition to do so. The Eiffel Tower became a blaze of light capped with a tricolour searchlight that extended its beam across half of Paris.

The telegraph section displayed the new multiplexing techniques with two or four messages sent down the same line, in use in America and Britain (where royalties to the Edison company were paid). An automatic telegraph, the phonoplex, displayed the transmission of images over ordinary lines, similar to the Bain invention, described earlier, with the driving motors at each end kept in synchronism. A Morse transmission set, using perforated paper tape, was shown, said to be capable of 720 words/min. Telegraphy from moving trains was demonstrated with short-range wireless transmission to wires at the trackside carrying an electrical signal corresponding to the Morse code. This used a simple apparatus of a battery, induction coil, vibrator, Morse key and 'phonetic' receiver. An aerial, described as 'the condensers', was carried on the train roof. Other telegraphy innovations included private line telegraphy, embossing telegraph, harmonic telegraph, and a stock quotation printer. A separate instrument section was appended which contained several versions of Wheatstone's bridge and other telegraph testing instruments.

In addition to the Edison generators shown at the Exposition, the French Thomson–Houston company showed their huge 80 hp DC generator, which provided a current of 125 A at 500 V. One of these generators had recently been installed in the United States for a tram installation at Bangor, Maine [47]. It was stated that this was the largest generator ever built by Thomson–Houston at that time. They also exhibited AC generators based on a design by Westinghouse. The Thomson–Houston electric welder was also exhibited, giving a 2 h performance each day, welding 6 cm steel bars.

All of these exhibits and other appropriate mechanical structures were housed inside a huge Galerie des Machines, which reached a height of 45 m and covered 6 ha. The 116 m width of the building was spanned by a glazed roof and unsupported in the centre, making this building the most striking feature of the Exposition. Its design by the French engineer Cotamin was remarkable for the time. To cover the largest possible area of ground with the smallest number of roof supports Cotamin was to balance his gigantic roof on huge hinged supports with pins at the apex of each arch to allow the great structure to accommodate the necessary amount of structural movement within itself (Figure 9.8). A striking internal construction was an electrically operated platform which could carry 200 visitors down the length of

Figure 9.8   Galerie des Machines, Paris, 1889
(Illustrated London News, 1889)

building at a height of 6.7 m, providing a spectacular view of the interior.

9.10 Chicago Columbian World Exhibition, 1893

A special electrical exhibition formed part of the Columbian World Exhibition at Chicago in 1893. This Exhibition was formed to celebrate the 400th anniversary of the discovery of the New World by Columbus in 1493. It was held at a site in Jackson Park on the shores of Lake Michigan.

A great structure of 2.23 ha was set aside in the grounds of the Exhibition for electrical exhibits. This was the electricity building, lit by 120 000 lamps and dominated by a great statue of Benjamin Franklin in the grounds outside the building.

As always in planning an international exhibition of this size, it became necessary to consider transport within the grounds. Two systems were in use. An intramural electric railway, consisting of three open trucks and a driving car circulated the site on an overhead structure, taking about 20 min to complete the journey at a maximum speed of 19 km/h. This was one of the first light railways to use a third rail power source. The driving car was equipped with four motors, one to each axle. Control of the motors was carried out by first connecting them in series to provide maximum torque for start-up, then in series as two parallel connected sets and finally all in parallel for constant running [48]. The system was installed by General Electric and its generating plant formed one of the most popular exhibits. It comprised one 200 kW, one 500 kW and one 1500 kW railway generator; all were DC machines and directly coupled to three steam engines, the largest being a great 2000 hp cross-compound Reynolds–Corliss engine [49].

The second transport system was extremely innovative, being the first continuous moving platform to carry passengers over any considerable distance. It was a multiple-platform, multiple-speed device, as shown in Figure 9.9, using two contiguous platforms—an inner one moving at a speed of 4.8 km/h and an outer one containing seats travelling at 9.7 km/h. Handrails were fixed to the inner platform to assist stepping from one platform to the other [50, 51]. It was laid out in a great ellipse having a total length of 1310 m. Seats were provided for 4212 people and the scheme had a total capacity of over 31 000 seated passengers/hour. Every 35th platform carried two 11 kW motors, and a total driving power of up to 200 kW was

Figure 9.9 Moving platform at the Columbian World Exhibition, Chicago, 1893 (Reproduced with permission from Scientific American, 1892)

demanded. Although various moving platform systems had earlier been set up for short distances, this was the first practical system to be used. It was designed by J.L. Silsbee and M.E. Schmidt and, ten years after the Exposition, Schmidt was to receive the Franklin Institute Award for this work. An almost identical system was set up at the Berlin Industrial Exhibition of 1896.

The largest foreign participants within the building were France and Germany each with $2140 \, m^2$. Other major displays were by Britain, Spain, Cuba, Brazil and Canada. Some 86 foreign nations, colonies and principalities took part, the largest number ever to contribute to a single international exhibition.

The British exhibits, although not large due to the punitive McKinley tariff existing at this time for the import of foreign goods, did include a sizable contribution from the General Electric Co. (GEC) of Manchester and London. This consisted of domestic and industrial components for the electrical industry (GEC at that time were attempting to open a branch of their company in the United States). The domestic products exhibited included electric cooking utensils of all kinds and, as claimed by a reporter of the time,

> '*a device of (great) interest to our lady readers ... which is nothing less than an electrical device for heating curling irons*' [52].

As at Frankfurt, a separate hall contained the large power plant required to drive the electrical exhibits and services. Edison and Westinghouse were rival contenders for lighting and power at the Exposition. They had clashed earlier in 'the battle of the currents' on the alternative merits of DC (Edison) and AC (Westinghouse) for nationwide transmission standards within the United States. Alternating current was eventually adopted despite earlier legislative attempts, initiated by Edison, to ban the use of AC supply on the grounds of safety.[3] By 1900 more than 1 billion dollars had been invested in electrical installations in the United States, all powered by alternating current. The decision to back Westinghouse by the 1893 organising committee ensured that his company provided all the lamps, transformers, switchboards, cables, and most of the dynamos and power stations required for the site [12]. It was by far the largest lighting contract that had ever been awarded in the United States. A few years later the Westinghouse company received a contract for the

---

[3]One of the 'dirty tricks' employed by the opponents of AC was to point to the 'electric chair' (manufactured by Westinghouse and sold to the New York State government) as evidence of the 'unsafe' nature of alternating currents

massive Niagara Falls generating plant, which was celebrated at the Buffalo Exhibition of 1901, discussed in the following Chapter.

The lighting power plant installed by Westinghouse was an AC system including a number of Westinghouse dynamos with a collective power of 12 000 hp and a Siemens plant of 1500 hp. Some of these are shown in Figure 9.10. The lighting demands were immense: 8000 arc lamps each of 2000 cp, 130 000 incandescent lamps each of 16 cp, and searchlights in tall towers. The dynamo switchboard required to control this installation was over 12 m in length. Also driven by this power plant were the electric lifts which were a feature in all the buildings of the Exhibition. General Electric were responsible for the operation and lighting of the many illuminated fountains that appeared in the grounds. Thirty-eight 90 A projector lamps using

Figure 9.10   The Westinghouse Power Plant at the Columbian World Exhibition, Chicago, 1893
*(Electrical World of New York, 1893, **21**)*

parabolic reflectors focused on the streams of water that issued from nearly 400 apertures. They were driven by four 175 kW Edison DC generators which, after the close of each evening's operation, were used to charge the batteries of a fleet of 50 electric launches that plied the lagoons during the day.

The Westinghouse company also exhibited a number of its other products. One of these was most important to the public distribution of electric power in the United States (and indeed in Europe). This was a convenient way of measuring the amount of alternating current consumed by the customer. The DC systems of Edison used the principle of electrolysis, having no application in an AC system.[4] One of Westinghouse's engineers, O.B. Shallenberger, was the first to devise, in 1888, an electromechanical device for AC power measurement, based on the rotation of a conductor in a moving magnetic field (effectively a small electric motor), a method which is in universal use today for measuring energy (or current in a constant voltage circuit). Also shown by Westinghouse were a number of DC generators designed for railway supply ranging in capacity up to a 600 kW unit directly coupled to a low-speed steam engine and a number of rotary converters, one of which was rated at 500 hp and designed to provide 500 V DC for railway service from an AC input. A number of the new Tesla constant-speed AC induction motors were also exhibited (Westinghouse had just acquired the Tesla patent rights to his invention which had been announced in 1888 [53]).

In the United States the impetus for the adoption of electricity for heating and cooking in the home began with the 1893 Chicago exhibition [54]. At this time the country was beginning to install electricity generating stations and had few gasworks. This may be contrasted with Britain, where over 1000 gasworks were in operation at this time and much use was made of gas for domestic purposes. At the Exhibition, apart from its use for lighting, domestic electrical exhibits included heating devices, hot plates, flat irons, electric fans, cooking utensils, electric bells, bed warmers, radiators and refrigeration equipment [55]. All the possible industrial uses of electricity were exhibited: ovens, furnaces, electrometallurgy, electro-chemistry, forging, stamping, drilling, cutting, welding, the electric telegraph and telephone, the phonograph, railway signalling and even

---

[4]Edison's method of measuring DC electrical energy depended on electrolytic deposition. When a current is caused to flow through a solution of zinc sulphate, into which two zinc places are immersed, zinc is dissolved off one plate and deposited on the other. The rate of deposition is 1.213 g/A/h, which translates into kWh with a fixed supply voltage

an electric chair! Elisha Grey, the American inventor, who very nearly produced a working telephone before Bell, demonstrated his telautograph which reproduced the sender's handwriting as he was actually writing on the sending instrument. This was an improved version of Bakewell's copying telegraph shown at the 1851 exhibition in London (see Chap. 5). The number of electrical exhibits exceeded that of all earlier exhibitions, with over 500 exhibits shown on the Bell Telephone stand alone.

Electrical artifacts were also to be found situated in the grounds: a 1500 hp electric locomotive on loan from the Baltimore & Ohio railway (the first main line electric railway), and numerous electric launches to be seen in the lagoons.

Finally, mention must be made of an engineering achievement which found a place in a number of later exhibitions and fairs—this was the first gigantic 'wheel of pleasure', constructed by George Ferris, a United States engineer, in 1893 for the Chicago Exposition, a giant mechanical marvel which transported 2000 people at a time 76 m into the air (Figure 9.11). It was later moved to the St Louis exposition in 1904, but by this time others had built similar 'big wheels', and in Europe the British civil engineer H.C. Booth (he invented the vacuum cleaner in 1901), erected one in the grounds of the old Earls Court exhibition and another at the Prater in Vienna, which still turns today.

Held at the same time as the Columbian Exhibition an International Electrical Congress was taking place in Chicago. This continued the work on electrical standards, commenced at the previous Congress in Paris in 1881, and issued a series of recommendations which were given legal force in Britain in 1894. These included definitions of the international ohm, ampere, volt, coulomb, farad, joule, watt and henry which have been universally accepted and remain in use today, based on the metre, kilogramme and second (MKS) system.

A few years later in 1896, New York was to present its own Electrical Exhibition. This was very much a vehicle for Edison's inventions and the prodigious output of his factories, by then located in several States. It was notable for the first public display of an X-ray tube, which Edison employed in conjunction with a fluorescent target material (calcium tungstate) to display the effects of X-rays through an intervening body. He called his arrangement a fluoroscope and was able to give his demonstrations only a year after Roentgen's discovery in 1895. Four sets of apparatus were exhibited by which visitors, '... *were able to inspect their own anatomy*'. By the end of the Exhibition between

*Figure 9.11   The Fairground Wheel of George Ferris, first shown at the Columbian
World Exhibition, Chicago, 1893*
(Chicago Historical Society, 1893)

3000 and 4000 people had taken advantage of this opportunity,
causing the editor of the *Pall Mall Gazette* to comment,

> *'It is now said, we hope untruly, that Mr Edison has discovered a
> substance ... potential ... to the X-ray, the consequence of which
> appears to be that you can see other people's bones with the naked eye*

... On the revolting indecency of this there is no need to dwell, but what we seriously put before the attention of the Government is that the moment ... (this) comes into anything like general use it will call for legislative restraint of the severest kind' [56].

9.11 References

1 PREECE, W.H.: 'St. Pancras Vestry Electricity Exhibition', *The Electrician*, 1891, **26**, pp. 610–612
2 'Electric lighting at the Paris Exhibition', *J. RSA*, 1881, **30**, pp. 98–107
3 'General Official Catalogue of the 1881 Paris Exhibition' (A Lahure, Paris, 1881)
4 BRIGHT, A.A. 'The electric lamp industry' (Macmillan, New York, 1949)
5 *The Electrician*, June 1882, **10**, p. 107
6 BRIGHT, and HUGHES: 'Report upon the International Exhibition of Electricity in Paris 1881', *J. Soc. Tel. Eng.*, 24 Nov. 1881, **X**, pp. 402–429
7 BOWERS, B.P.: 'The Electrical Exhibitions of 1881 and 1882'. IEE Weekend Meeting on the History of Electrical Engineering, Imperial College, London, 4–6 July 1980
8 CROMPTON, R.E.: 'Reminiscences' (Constable, 1928), pp. 98–99
9 BOWERS, B.: 'Edison and early electrical engineering in Britain', *in* HOLLISTER-SHORT and JAMES (Eds): 'History of technology, vol. 13' (Mansell, London, 1991), pp. 168–180
10 HARRIS, J.E.: *The Times*, 3 Dec. 1994. Letter
11 BOWERS, B.: 'History of electric light and power' (Peter Peregrinus, London, 1982), p. 124
12 OLIVER, J.W.: 'History of American technology' (Ronald Press, New York, 1956)
13 'The Electrical Exhibition at Paris—Hopkinson's electrical hoist', *The Electrician*, 19 Nov. 1881, **8**, p. 10
14 GOETZELER, H.: '100 Jahre elektrischen Lichtbogenofen', *Elektrotech. Z. ETZ*, 1979, **100**, p. 558
15 THOMPSON, S.P.: 'The life of Lord Kelvin, vol. 1' (Macmillan & Co., London, 1910), p. 575
16 HEAP, D.P.: 'The Paris Electrical Exhibition of 1881' (Dept. of War, Washington, 1884)
17 Editorial: 'The Electrical Exhibition at Paris', *The Electrician*, 8 Dec. 1881, **VII**, pp. 40–41
18 See *The Electrician*, 15 October 1881, p. 344; *Saturday Review*, 11 February 1882; *The Athenaeum*, 22 April 1882
19 FLEMING, J.A.: 'Memories of a scientific life' (Marshall Escott Ltd., London, 1934)
20 Editorial: 'Crystal Palace 1882', *The Electrician*, 1882, **8**, pp. 216–217
21 *Engineering*, 1883, **35**, p. 205
22 HOLMES, A.B.: 'Practical electric lighting' (E & F.N. Spon, London, 1887, 3rd edn.)
23 PREECE, W.H.: 'The Munich Electrical Exhibition 1882', *J. Soc. Telegraph Eng.*, 1882, **11**, pp. 517–545
24 'Deprez's transmission with direct current', *The Electrician*, 1885, **10**, p. 40
25 'Electric railway at Vienna Electrical Exhibition', *Electr. World*, 5 July 1884, **1** (1), p. 1
26 PASSER, H.C.: 'The electrical manufacturers 1875–1900' (Harvard University Press, Cambridge, Mass., 1953)
27 *The Electrician*, 27 November 1891, **28**. Editorial
28 GIBSON, J.M.: 'The International Exhibition of 1884: a landmark for the electrical engineer', *IEEE Trans. Educ.*, 1980, **E-23** (3), pp. 169–176
29 'The forthcoming exhibition at Philadelphia', *The Electrician*, 1883, **11**, pp. 483
30 POMERANTZ, M.A.: 'Journal of the Franklin Institute: for the diffusion of scientific knowledge', *J. Franklin Inst.*, 1966, p. 301 (150th year celebration journal)
31 MACLAREN, M.: 'The rise of the electrical industry during the nineteenth century' (Princeton University Press, Princeton, 1943)
32 'Weston's plating generator', *Telegraph. J. Electr. Rev.*, 1878, **6**, p. 159

33 'Report of the Electrical Conference at Philadelphia in September 1884' (Government Printing Office, Washington, 1886)
34 *J. Franklin Inst.*, 1885–1886, Suppl., Sept., **120**, pp. 53–55; Nov., p. 8; **121**, pp. 149–154; **122**, Dec. pp. 448–460
35 GUILBERT, C.F.: 'The generators of electricity at the Paris Exposition of 1900' (C. Naud, Paris, 1902)
36 'Philadelphia Exposition', *Electr. Rev.*, 7 Feb. 1885, **6**, p. 4
37 LANDES, D.: 'The unbound Prometheus' (Cambridge University Press, 1969) p. 151
38 'The Frankfurt International Electrical Exhibition', *The Electrician*, 1891, **26**, pp. 686–687
39 'Allgemeiner Bericht über die internationale Elektrotechnische Austellung in Frankfurt am Main 1891, vol. 2' (D. Sauerland Verlag, 1851)
40 SINGER, HOLMYARD, HALL and WILLIAMS: 'A history of technology, vol. V, 1850–1900' (Oxford University Press, 1958), pp. 230–233
41 'The Crystal Palace Exhibition', *The Electrician*, 1892, **28**, pp. 253, 458–459
42 WILLIAMS, T.: 'A history of technology 1900–1950', Oxford University Press, 1978, **7**, pp. 1132–1133
43 CORLEY, T.A.B.: 'Domestic electrical appliances' (Cape, London, 1960)
44 MAX DE NANSOUTY: 'L'anné industrielle' (Paris, 1887), p. 14
45 'The second Crystal Palace Exhibition', *The Electrician*, 1892, **28**, p. 25
46 STASSNY, W., and ROSSETTI, E.: 'L'Exposition Universal de Paris' (American Commission, US Govt., Paris, 1889)
47 'The Electric Railway at Bangor, Maine', *Electr. World New York*, July 1889, **14** (1), p. 10
48 'Polyphase system at the Chicago Fair', *Electr. World New York*, 1893, **21**, pp. 10, 91, 374
49 'General Electric Company's Intra-mural Railway at Chicago', *Electr. World New York*, 1893, **21**, p. 335
50 'The travelling sidewalk at the World's Columbian Exposition', *Sci. Am.*, 1892, **66** (3), p. 31
51 'The moving sidewalk in Chicago', *The Electrician*, 1893, **3**, pp. 803, 633
52 'Electricity at the World's Fair', *Electr. World New York*, 1893, **21** (23), p. 426
53 TESLA, N.: 'A new system of alternating motors and transformers', *AIEE*, 1888, **5**, p. 308
54 GIEDION, S.: 'Mechanization takes command' (Oxford University Press, New York, 1948)
55 'Official guide to the world's Columbian Exhibition of 1893' (Columbian Guide Co., Chicago, 1893)
56 ALLISTER, R.: 'Friese-Greene; close up of an inventor' (N.Y. Arno Press, 1948), p. 77

Chapter 10
Dominance of technology

10.1 World fairs and international competition

With the turn of the century the technological content of large international exhibitions increased and much more space was set aside for electrical and other specialist technologies. Although architecture and sculpture continued to play an important part in the layout of the exhibits, in buildings and in display within the grounds, these were more directly associated with the particular technology being exhibited or, in later exhibitions, the theme of the exhibition. This was also the case for art exhibits, such as murals, which were designed to emphasise or bring out some aspect of the technology concerned. The separate display of painting and objets d'art, for so long a feature of public exhibitions in the second half of the 19th century, was now confined to separate art exhibitions and ceased to compete for attention within the same grounds as the educational, engineering and technology exhibits.

One result of this was that the new, technological, but still public, international exhibitions became larger and more competitive. Despite the participation of many different countries at a given exhibition, each having its own building, pavilion or display area, the host countries providing venues for these events vied with each other on a worldwide stage in order to attract visitors from abroad. The international exhibition thus became not only a public informative and entertaining occasion, but a potentially rewarding event as well, with its 'success' or 'failure' judged by the balance sheet. This was to become marked in the early years of the twentieth century with the New York exhibition of 1939/40; although pronounced by participants and public alike a resounding success, it was considered a financial failure.

This competitive trend is seen most clearly in the Paris 1900 exposition held at the Esplanade des Invalides, the Champ de Mars, the Trocadero and neighbouring quays. The French international exhibitions were not only larger than those promoted elsewhere but were projections of the government itself and were lavish in their presentation. The question of commercial success did not, at that time, enter in the planning for these events. The rationale for this view was stated by Baron Taylor when he created the Union Centrale des Arts Decoratifs (L'UCAD), supported by the French administration, and founded after the British Great Exhibition of 1851 to '...*ensure a continued place for the promotion of French art in the world market.*' This soon developed into an organisation for the promotion of international exhibitions, either in France or abroad where a substantial French contribution is made. In more recent times this organisation and its influence is seen to be secondary to the Bureau International des Expositions (BIE), and L'UCAD is now almost entirely concerned with its new role of industrial design. (The BIE is considered in detail in Chap. 13.)

Technology was also much in evidence at the smaller exhibitions at the beginning of the century, all taking advantage of the widespread availability of electrical power. This was seen at the 1907 international exhibition in Dublin, which included a Palace of Industry and a Palace of Mechanical Arts with a good display of electrical appliances of interest to the public. An electric tram-car service had just been installed in Dublin which gave easy access to the exhibition site. Glasgow held its extraordinarily successful International Exhibition in 1911 and attracted twice the number of visitors as the 1851 Exhibition. Brussels built a huge Galerie des Machines for its 1910 international exhibition and included electricity as a separate category of exhibits. The Turin Industry and Labour exhibition of 1911 attracted over 4 million visitors to its display of engines in motion and electricity including an electric elevated railway. Even a small country such as New Zealand was assiduous in promoting exhibitions, competing in numbers with France in sponsorship of exhibitions towards the end of the 19th and for most of the 20th century. Twenty were held between 1865 and 1926, and although all but five were national events concerned with agriculture, the international exhibitions held in Christchurch (1906–1907) and Dunedin (1925–1926) fully embraced the need to express the countries' technical progress. The Christchurch exhibition contained a machinery hall and attracted 2 million visitors.

It was, however, the Paris Exposition of 1900 which set a record in size and technical magnificence in Europe for over half a century—until the Brussels Exposition of 1958, considered in Chap. 13.

10.2 The Paris International Exposition, 1900

The Paris 1900 Exposition opened for seven months and attracted almost 50 million visitors, a figure rivalled only by the earlier French Exhibition of 1889. It occupied a site larger than any other previous Paris exposition with its main ground of 113 ha and an annex at the Bois de Vincennes of 111 ha. Apart from outstanding new buildings it incorporated existing buildings remaining from the 1878 and 1889 Expositions, such as the Galerie des Machines and the Eiffel Tower within the Champ de Mars. Almost 100 French and 75 foreign pavilions were distributed over the grounds. Although vast numbers of visitors attended the Exposition daily the huge size of the exhibition site led to not a little confusion, which was not helped by the method of classification used for the exhibits. A large number of exhibits, some 83 000 in all, were distributed amongst 40 nations and classified by subject groups and classes rather than by nation of origin. In addition, numerous foreign pavilions and attractions had their own classification arrangements, which added to the complexity.

The electrical exhibits were of great interest to the professional visitor to the Exposition as well as to the general public. Sir William Preece reported that members of the American Institute of Electrical Engineers (AIEE) and the British Institution of Electrical Engineers (IEE) *'foregathered in London and chartered a special steamer to take them to France to visit the Exposition'* [1]. Most of the electrical exhibits were housed in the Palais de l'Industrie (Figure 10.1), which repeated the spectacle of electric illumination shown at the 1881 Exhibition. The facade was lit by the changing lights of its 5000 multicoloured incandescent lamps. At the top a Spirit of Electricity, driving a chariot drawn by hippogryphs, projected showers of coloured flames. An orchestra played a piece entitled, 'Ode to Electricity'. Visitors crowded into the building to watch the steam-driven dynamos power the exposition machines, a spectacle full of noise and light, described by a journalist, Paul Morand, as *'L'electricite, c'est le fleau, c'est la religion de 1900'* [2].

The generator equipment required for these and other applications of electricity was immense and shows clearly the state of design and construction of dynamos at the turn of the century. Full details of these were later compiled by M.Guilbert, an electrical engineer, in his

Figure 10.1   Le Palais de l'Electricité, Paris, 1900 (Bibliotheque Nationale)

book entitled, *'The Generation of Electricity at the Paris Exhibition of 1900'* [3]. In its 700+ pages of detailed description, plans and tables, the author presents a valuable account of generator technology not seen since the publication of a similar report on electrical generators exhibited at the Frankfurt Electrotechnology Exhibition of 1891, noted earlier. A number of performance graphs were included in Guilbert's reports which were obtained with the aid of André Blondel's electromagnetic oscillograph, a new device which had only just become available in 1893 [4].

The generators were classified into groups:

Part 1: Alternators
    alternators with solid field poles
    alternators with laminated field poles
    inductive alternators and compounded field alternators.
Part 2: Rotary transformers
    motor generators
    converters
    rectifiers.
Part 3: Direct current generators
    direct current generators with steel fields
    direct current generators with composite fields
    direct current generators with cast iron fields.

Each group in the machinery hall comprised generators contributed by several countries at the forefront of design at that time: Germany, Switzerland, France, Sweden, Austria, Italy, Britain, Belgium and the United States. The total available generator capacity was 42.7 MVA.

The Paris 1900 Exposition also installed a moving pavement or 'Trottoir Roulant' for the convenience of visitors. This was electrically operated and caused considerable interest in the visiting public. Although not the first to be seen at an exhibition (the first was installed at the Columbian World Exhibition at Chicago in 1893), it was certainly the longest. It consisted of three concentric platforms, the first moving at a speed of 5.3 km/h, the second at 10.8 km/h and a third at 16.2 km/h. The fastest platform was fitted with seats (Figure 10.2). A Paris newspaper noted,

> '... *a railway without weight, noise or smoke; without cinders, smells or jars; where crowding and waiting is unknown; on which passengers cannot be knocked down by cars or have their legs cut off by wheels'* [5].

Figure 10.2   *'Trottoir Roulant' shown at the Paris 1900 Exposition*
(*Illustrated London News*, 1900)

It was a much needed convenience and, together with an electric train which covered part of the site, enabled the huge size of the grounds to be somewhat mitigated. To drive the moving pavement, 172 drive units (Alliot DC motors), each requiring 3.7 kW, were distributed around the circuit. The moving pavement was a major feat of engineering [5]. For most of its 3.36 km length it was located along and across streets, when it used a viaduct structure. It had nine stations each reached by a moving ramp—a structure which saw its first public appearance at the Paris Exposition. Twenty-seven of these moving ramps were installed throughout the Exposition. The Trottoir Roulant was supplemented by a single track electric railway circulating in the opposite direction, but following the same route, consisting of eight trains, each carrying 250 passengers. The total capacity of both systems was 50 000 passengers per hour, and by the close of the Exposition the pavement had carried 6.7 million passengers and the railway 2.6 million.

Much of the electrical equipment, exhibited earlier at the 1881 and 1889 Paris Expositions, was seen at the 1900 Exposition. By 1900 the development of industrial electric motors was sufficiently advanced for them to be used to drive workshop tools directly, instead of the installation of overhead shafting and a forest of belting leading down to the machines. At the exposition many such electrically driven tools for drilling, milling, turning, etc. were shown. Electric vehicles were prominent. Some 50 were to be seen, many of American manufacture, where a large exhibition of such vehicles had been seen at Madison Square Gardens in New York a few months earlier [6]. Some of the French vehicles incorporated novel features, such as those from Ch. Mildé Fils & Co which incorporated a differential motor having two armatures revolving in a single field. The armatures revolve independently, each engaging directly with one of the drive wheel gears, thus dispensing with the use of a set of differential gear wheels. A German company, the Gesellschaft für Verkehrsunternehmungen of Berlin, showed an electric omnibus accommodating 17 passengers, carrying an overhead charging rail, rather like a tram, which came into action when the vehicle ran into a bus station [7].

Amongst the exhibits was Valdemar Poulsen's wire magnetic recorder, capable of recording speech permanently on a steel wire and known as 'the microphonograph'. At a later exhibition in 1910 at Brussels the wire recorder of Ernst Rühmer was exhibited for the first time. He was the first to apply the modulation principles of radio technique to the recording of speech on wire in 1909 [8].

Considerable interest was shown in the numerous presentations of the moving image. Seventeen projection booths were in operation and many dioramas, panoramas, phenakitoscopes, stroboscopes, zoetropes, etc. were to be seen about the Exposition. A major demonstration was, however, the first demonstration of the Cinérama. This was designed by Raol Grimoin–Sanson, a French inventor of early motion-picture equipment [9]. He used ten 70 mm projectors synchronised to a central motor and producing a picture over an angle of 330 degrees projected on to the inside walls of a vast marquee. This was arranged so that the viewers stood on an elevated platform in the centre, intended to represent the nacelle of a balloon, thus giving the impression of a view of earth as seen from a great height. Unfortunately a combination of overheating and inflammable film stock resulted in the abandonment of the project after only three days. However, the idea had been presented and this remains the first public performance of a technique which was to be seen in many subsequent exhibitions, most spectacularly in the IMAX wide angle film presentation at the American Knoxville Energy Fair of 1962.

A new feature, which was soon to be repeated at other international exhibitions, was a Salon d'honneur located in the Chateau d'Eau, where homage was paid to French and foreign engineers with displays showing their principal inventions. Amongst those named were Tesla, Edison, Wehnelt, Marconi and Dussaud.

10.3 Glasgow International Exhibition, 1901

This followed Glasgow's successful first international exhibition in 1888 and Glasgow's 1880 Electrical Exhibition, considered in the previous Chapters, and similarly devoted considerable space to heavy industry and the burgeoning electrical industry. It was held to celebrate the 50th anniversary of the Great Exhibition in London of 1851 and also, incidentally, the opening of Glasgow's new City Museum and Art Gallery which formed part of the Exhibition. The Exhibition was held on a 30.4 ha site in Kelvingrove Road with significant buildings given to an industry hall (28 000 m^2) and a machinery hall (15 000 m^2). The industry hall was outlined in electric light and crowned with a golden dome and occupied a large portion of the Exhibition space. The electric generation equipment in the machinery hall followed a similar arrangement to that seen at the 1888 Exhibition. Several nations had their own pavilions, including Canada, Ireland, Japan and Russia.

Exhibits were arranged in eight classes covering raw materials, industrial manufacture, machinery & power, transportation, marine engineering, lighting & heating, science, music and sport. The organisers arranged a number of examples for comparison of various systems of illumination by means of electricity, gas, high-pressure gas, acetylene gas and high-pressure oil. Much use was made of 10 A Brockie–Pell arc lamps in the industrial hall and grand avenue; almost

1000 were used, together with over 5000 incandescent lamps. Within the machinery hall the *Glasgow Herald* set up a substantial electrically driven Hoe printing press, producing its newspaper during the 6 months' run of the exhibition at speeds of about 24 000 pages/h [10].

Electric launches were in operation on the river Kelvin, and to reach the site visitors could travel on the 116 km of double tramway track which had been completed in time for the Exhibition. To provide power for this project the generators were, surprisingly, American, from the Westinghouse Company of Milwaukee, but the new heavy tram-cars were manufactured in Glasgow.

Unlike many international exhibitions at the time the Glasgow event produced a financial surplus and obtained a total attendance of 11.5 million over the 6 months it was open.

10.4 Franco–British Exhibition, London, 1908

This was held in the White City, Shepherds Bush, and, although limited to the products of two nations and their empires, was very successful and the largest exhibition ever held in London until the Wembley Empire Exhibition of 1924. It covered 56.7 ha and attracted 8.3 million visitors. It took place at the height of the 'Entente Cordiale' and was visited by the French President during the first foreign journey undertaken by the holder of that office.

The idea of holding the Exhibition was suggested by Imre Kiralfy, who eventually planned and designed the entire Exhibition. Imre Kiralfy was at that time in the middle of his career as a writer and organiser of exhibitions. He had already acted as Director-General of six earlier exhibitions at Earls Court from 1895 to 1903, including the India Exhibition of 1895 and the Victoria Exhibition of 1897. He went on to direct the commissioning of five other exhibitions at Shepherds Bush and on the continent, of which the Coronation Exhibition of 1911 was the most extensive. His work at the 1908 Exhibition included the conception and design of the White City Stadium erected for the 1908 Olympic Games and capable of containing 100 000 spectators.

The site contained seven spacious exhibition halls, each covering an area of 2600 m², and a vast machinery hall containing many of the electrical exhibits. The Science Committee responsible for the technical aspects of the exhibition was chaired by Sir Norman Lockyer and included several names well known in the electrical engineering world: Sir William Preece, Sylvanus P. Thompson, R.E. Crompton,

A. Siemens, Professor Rutherford and Dr. R.T. Glazebrook. A large section of the Exhibition was devoted to education, from elementary school to university, in response to the current interest in Britain's standing *vis-à-vis* the continent at this time, and was the most extensive attempt to portray an overall picture of education ever mounted in a British exhibition.

The machinery hall housed a display of various coal seams, a mining section and a large installation of pumping machinery. It also contained a 2000 hp steam turbine plant directly coupled to an electric generator providing current for illumination of the Exhibition and its gardens. This was supplemented with a 1000 hp vertical gas engine which drove further generators for the lighting installation. The entire electrical plant was supplied and managed by the combined Electric Supply Companies of London.

At this time public electricity supply installations did not command a large distribution area, and many public buildings and large country houses installed their own private systems. A market existed for the supply of lighting sets for this purpose, and the 1908 Exhibition was one of the first public exhibitions to demonstrate this need. Several firms showed their lighting sets for country houses, including Hubert & How of Wimbledon, British Thompson Houston of Rugby and, in France, Société Gramme and Appareillage Electrique Grivolas of Paris. Many of the exhibitors in the machinery hall displayed their range of light fittings and control equipment. Prominent amongst these were the Edison and Swan United Electric Light Company before they changed their designation to Ediswan Ltd, and smaller electric lamp companies such as the Stearn Company, all busily engaged in competition with Edison and Swan. Several manufacturers displayed electric motors used in machines for a variety of purposes; woodworking, pumps, organ bellows and, presumably for export, electric punkahs and fans, together with a range of heaters, cookers and flat irons. Other technical areas included electrochemistry and, as part of the French exhibits, a range of electrofurnaces, aluminium smelting machines and applications of electrolysis. The use of coal-dust for microphones and telephone equipment was demonstrated by a société anonyme, Le Carbone of Paris.

10.5 Three large American exhibitions, 1900–1915

The industrial power and innovation provided by the American people began to dominate the role of international exhibitions at the beginning of the 20th century. That this was likely to be the outcome had already been seen in the Chicago World Fair of 1893 and the American contributions to the large French Expositions of 1881, 1889 and 1900. The series of large American Exhibitions in Buffalo 1901, St. Louis 1904 and San Francisco 1915 established this trend, which continued unbroken until challanged by the Far East exhibitions towards the end of the century. New York was not to contribute to the American exhibition scene until after World War II, although an abortive attempt was made to stage a World Fair in the Bronx in 1918. This proved a failure but was noticeable for the public showing of the first practical submarine, designed by John P. Holland, an American citizen.

10.5.1 Pan-American Exposition, Buffalo, 1901

The Pan-American Exposition held at Buffalo in 1901 was particularly well placed for electrical demonstrations at its Palace of Electricity due to its proximity to the massive Niagara Falls Generating Station. This had been opened a few years earlier in April 1895 and, to some extent, the Exposition was regarded as a celebration of that event for the town of Buffalo. During its many years of planning and execution (the project commenced in 1886) the alternative arguments for DC or AC transmission were being pursued with the International Niagara Commission in London. The success of the Westinghouse AC installation at the 1893 Chicago Exhibition has been cited as a significant factor in the ensuing appointment of the Westinghouse Co. as the major contractor for the Niagara Falls' installation in May 1893. This decision was reached despite the intervention of Lord Kelvin, who cabled from London the succinct advice, *'Trust you avoid gigantic mistake of alternating current'* [11].

The installation initially produced a turbine power of 5000 hp but this grew to 20 000 hp by 1897, when ten turbines delivered a two-phase supply at 2200 V at the low frequency of 25 Hz [12]. By 1904 the station achieved its design maximum of 100 000 hp with more generators added. At the time of the Buffalo Exposition its Westinghouse AC generators were used to power over 200 000 lights installed in a huge 119 m electric tower, in street illumination throughout the grounds, in outlining every exhibition hall, in lavish interior lighting and in providing power to the electrical exhibits [13]. The Omaha Exhibition of 1898 had pioneered the technique of lighting the grounds entirely by incandescent lamps, and the Buffalo Exhibition profited by this experience through the use of large numbers of 'Omaha posts', a 2.5 m post surmounted by a cluster of twenty 16 cp (16.3 cd) incandescent lamps.

At its opening the Exposition was declared to be '*the electrical marvel of the new century*'. Power was transferred to the Exhibition site at 11 000 V, and it was at this voltage that the lighting engineers chose to operate an immense 'water dimmer' to enable the lights to be gradually switched on. It consisted of three separate tanks of water, one for each phase, and each measuring 2.4 m × 1 m × 1 m, having a cojoint capacity of over 6000 litres. The bottom of the tanks formed part of the main circuit, whilst an iron blade 2 m long, acting as the outlet terminal, was gradually lowered into the water, a process causing a vicious arcing to the surface of the water. This procedure formed part of the nightly entertainment for those privileged visitors admitted to the lighting control room [14]! General Electric set up a large model of its Schenectady works and surrounded this with numerous types of dynamos, alternators and motors. A complete alternating current supply system was exhibited by GE to provide current for arc lamps required on the site. The supply frequency from Niagara was at 25 Hz and too low for successful arc lighting, and a motor-alternator generator of 100 kW capacity was found necessary for this purpose. This was one of the last exhibitions to provide an extensive display of static electricity generation. These were large Wimshurst machines used by the medical fraternity in America for '*... cases of neuralgia, sciatica, rheumatism and kindred nervous and muscular disorders*'. They were also used for X-ray work, and a static machine exhibited by a New York company, '*... gave torrents of noisy sparks, 40 cm long*'. Demonstrations of its use connected to an X-ray machine were given from time to time at the Exhibition. Finally, another exhibit with a medical aspect was an incubator for premature babies, extravagantly praised by the *Cosmopolitan* magazine as '*one of the greatest inventions since the Chicago Fair eight years earlier*'.

In other ways, however, it was not a success [15]. The weather was poor and attendance much reduced. A serious disaster befell the exhibition half-way through its term. The assassination of President McKinley, whilst visiting the exhibition in September 1901, was a major catastrophe from which the Exposition never recovered, although it went on after a short closure to its scheduled end date of 2 November 1901.

10.5.2 St. Louis, 1904

The Louisiana Purchase International Exposition, held at St. Louis in 1904, commemorated the centennial of the land purchase in 1804 that brought Missouri into the United States, together with all the land from the Mississippi river to the Rockies. The area of the Exposition was larger than anything previously attempted in the United States (515 ha), and to achieve this major civil works had to be carried out in a heavily forested area , now known as Forest Park, a 40 min drive from downtown St. Louis. To facilitate transport into the city a new railroad was built and 450 new railway cars added to the rolling stock. Trains to the Exposition ran every 15 min. A 27 km electric intramural railway was also built to provide transport around the grounds. This was certainly needed—over 1500 separate constructions made up the Exhibition, including 15 'exhibition palaces' of major importance. Even with this assistance a journey around the Exhibition must have been exhausting, with a tramp of 14 km to see all there was to see in the largest pavilion, which was devoted entirely to agriculture!

The Exposition was opened remotely by President Theodore Roosevelt, 1100 km away in the White House, using a special gold telegraph key (the same key used to initiate the Chicago Exhibition of 1893).

Developments in aeronautics were prominent. Five of the early dirigibles were sent to St. Louis and a special airfield constructed for their use. The United States airship, *Californian Arrow,* made its first successful American dirigible flight from this airfield during the period of the Exposition.

The telegraph exhibits demonstrated, for the first time in the United States, wireless telegraphy between ground and air. At distances varying from 400 m to over 3 km, three men in a dirigible received 20 messages transmitted from a station within the grounds. Other wireless equipment was shown, together with a very early version of Lee de Forest's three electrode thermionic tube (the triode valve). This must have been one of the first displays of this new invention to the general public (Lee de Forest did not apply for a patent until 1906).

In general the electrical exhibits were grouped into five sections:

Group 67: contained machines for electrical power generation. An American exhibit was of a steam turbine driven alternator using a Curtis & Parson's machine, the result of Anglo–American co-operation

Group 68: was devoted to electrochemistry

Group 69: included electric lighting. General Electric exhibited a new form of lamp giving 'daylight colour' obtained by combining glow lamps with mercury vapour lamps

Group 70: showed exhibits in telegraphy and telephony, where Britain had a substantial presence. The British Post Office had a display of land telegraphy and Muirhead gave an exhibit of cable signalling and testing. The United States demonstrated an automatic telephone exchange, which was beginning to be installed in American towns

Group 71: exhibited various applications of electricity in measurement. From Britain were shown a Kelvin balance, Weston and cadmium cells, Crompton potentiometers, galvanometers and other measuring apparatus. The National Bureau of Standards gave a display of their current measurement techniques.

Demonstrations were also given in the British Section of Duddell's oscillograph, Röntgen ray and high frequency apparatus. In a British Government report on the Exposition the writer states,

> 'There was little inducement for European manufacturers to go to the trouble of sending machinery due to small prospect of finding or extending their markets in America and, as a consequence, only France and Japan made any exhibition of an engineering nature. ... With high tariffs almost the only opening is in the department of scientific instruments in which Great Britain and Germany are in advance of the United States'.

Certainly the electrical industry in the United States was far more progressive than in Great Britain at the time, due mainly to the superior natural resources and consequent demand for electrical plant in all parts of the country.

The 1904 Exposition was a financial success, unlike the preceding American exhibitions. Profits were used to build the Jefferson Memorial in Forest Park to house the records of the Exposition and the collections of the Missouri Historical Society.

10.5.3 San Francisco, 1915

The Panama–Pacific International Exposition at San Francisco in 1915 was initially planned by the city to celebrate the completion of the Panama Canal in 1914, but it was later agreed to share this celebration with San Diego, where preparations were also going ahead for a smaller exhibition in the same year, to be known as the Panama–Californian exposition. A more important theme at the time for San Francisco was to use the occasion of an extensive quasicommercial international exhibition to advertise the city's commercial recovery from the devastating earthquake and fire of 1906.

Although planned as an international event, foreign participation was limited due to the outbreak of the First World War. Neither Britain nor Germany took part, although France contributed a full-sized replica of its Palace of the Legion of Honour to house its exhibits. Perhaps influenced by recent large international expositions in the United States and France, the 257 ha site was arranged as a series of courtyards, each designed by a different architect, which made the Exhibition compact and easier to survey on foot. There were three major courts: in the centre, the court of the universe, comprising the palaces of transportation, manufacturers, arts and agriculture; to the east, the court of abundance, enclosed by the palaces of transportation, mines & metallurgy, varied industries and manufactures; to the west, the court of four seasons, enclosed by the palaces of agriculture, food products, education & social economy and the liberal arts. A number of vast domed buildings supplemented these exhibition areas. To the south of the courts rose an immense glass dome, larger than St Paul's in London, which contained the palace of horticulture, and a further domed building containing a concert hall and a palace of machinery. At the periphery of this interlocking set of courtyards were the national and state pavilions and the Exposition grounds.

The Exposition made extensive and almost exclusive use of indirect lighting both for internal and external illumination. The new tungsten filament incandescent lamp made its first appearance at a public exhibition and was used in large numbers. For floodlighting 370 searchlights and 500 projectors were strategically placed on roofs and other vantage points. This was a dynamic display with a *'squad of United States Marines changing coloured filters on a battery of 48 searchlights located on a pier in the bay'* [16]. This presented an hour-long programme, accompanied by fireworks, smoke bombs and maroons, as well as clouds of steam from a motionless railway locomotive. Show business was by now firmly entrenched in the repertoire of a modern international exhibition!

The transcontinental telephone connection across the United States had been completed by AT&T just before the exposition opened in January 1915. This was made a feature of the Edison exhibits with a working line to New York, over which reading of the New York newspaper headings took place and, slightly further to the east, microphones were placed so that the visitors to the San Francisco Exposition could hear the pounding of the waves on the Atlantic seaboard. General Electric showed a domestic exhibit of a model home which they termed 'the home electrical'. It contained a built-in vacuum cleaner with outlets in each room and an electric piano.

In the grounds a large working model of the Panama Canal was set up, covering 2 ha. A moving platform of seats transported spectators around the exhibit while they held telephone receivers to their ears to hear a recorded description of what they were seeing. The receivers were driven by a bank of 45 Edison phonographs, timed to coincide with the observed visual representation. This was one of the first audio recordings of information on exhibits to be available to the visiting public, and earned Thomas Edison the (only) Grand Prize awarded. Thereafter they were to become a feature of almost all subsequent world fairs and expositions.

A further innovation for international exhibitions was the 'special event' promoted nearly every day. A highlight of the exposition was Edison Day, when on the 21 October, the invention of the incandescent lamp was celebrated by a splendid banquet in which all the courses were cooked by electricity. Separate events featured the lives of Theodore Roosevelt, Luther Burbank and Thomas Edison. The American Association for the Advancement of Science (AAAS) held its first Pacific Coast meeting at this Exposition. The AAAS had been founded in 1848, modelled on the lines of the British Association (Chap. 3).

10.6 Exhibitions for technologists

Whilst in Chap. 13 the separation of exhibitions into those intended to interest the general public and those to promote trade will be discussed, there is a third category in which the latest techniques and discoveries are displayed for the benefit of those already skilled in these arts i.e. 'exhibitions for technologists'. They are presented by many professional bodies and scientific societies, and their number has increased since the last half of the nineteenth century. In these exhibitions the aim is to explain, educate and compare ideas and products and not necessarily to sell them (although an element of the trade show is generally present). There are very many in this category of exhibitions, many of them yearly events, and it would not be practicable to describe or even list all of them as has been done for international exhibitions in Table 2.1.

They present a valuable view on the development of technology at the time of presentation and often provide the first opportunity to show a particular invention or process to the technical public. Three of the earliest were the private viewings of the Society of Arts, those promoted by the Royal Institution and those of the Royal Society.

Others were (and still are) arranged by the various professional engineering institutes, the Institution of Civil Engineers, the Institution of Mechanical Engineers, and the Institution of Electrical Engineers, and are open to the public. An early example was an exhibition mounted by the Institute of Civil Engineers in 1851 to display a number of possible architectural designs for the forthcoming Great Exhibition.

A second example, 100 years later, ushered in the transistor era with a seminal exhibition mounted by the Institution of Electrical Engineers in connection with an International Convention on Transistor and Associated Semiconductor Devices held at Earls Court in May 1959, although some limited transistor applications had already been seen at the National Radio Exhibitions in the 1950s. This was, however, the first comprehensive exhibition to display all aspects of transistor development and provided a valuable 'snapshot' of the position of semiconductor development at a crucial time when commercial devices were just beginning to emerge from the laboratory to take their place in the industrial scene. At this exhibition 83 stands displayed transistor development from eight overseas countries as well as Great Britain [17]. Foremost amongst these were the British companies: Associated Electrical Industries (AEI), British Thomson Houston (BTH), and the General Electric Company (GEC). AEI demonstrated a working silicon crystal pulling furnace using electron bombardment heating, and showed how other diffusion processes can form the p-n junction of a solid-state transistor (by 1959 point-contact transistors and germanium devices were of less interest to the manufacturers, who were concerned to fabricate semi-conductor devices from silicon due to its superior performance). New possibilities of semiconductor devices were shown by BTH with their electroluminescent display panels and devices, making use of the Hall effect for current amplifiers and measurement purposes. Transistor-ised communication equipment and missile guidance equipment was shown by GEC. Philips were showing for the first time n-p-n transistors (the OC139 and OC140), a germanium device operating at the unheard of cutoff frequency of 3–4 MHz, and industrial power transistors working at a 6 A level. Standard Telephone and Cables exhibited their transistorised communications equipment for the telephone network. Several transistor digital computers were beginning to be available commercially following the success of the first British transistor computer, the Metrovic 950, shown at an Earls Court exhibition two years earlier. Ferranti displayed their prototype 'Sirius' desk-sized transistor computer, a new compact venture, although still requiring paper tape input media and with a storage capacity limited to a few thousand bytes, maintained in 20 magneto-striction delay lines. Dr William Shockley, one of the inventors of the transistor, addressed the Convention at its opening session on new developments in solid-state devices.

A similar opportunity was taken by the American Institute of Electrical Engineers (AIEE) to present a view of current technology at about the same time. This was held at the Electrical Engineering Exposition in 1962, the first since the Institute was founded in 1885 [18]. A basic concept was to give its members an opportunity of seeing instruments or apparatus discussed later in the 115 technical sessions of its Winter General Meeting which followed. As a consequence the display booths showed a wide coverage of the position of electrical engineering at the time, including automation, business machines, computers, measuring instruments, regulators, radar, systems engineering, machine tool applications, magnetic materials, power installations, servomechanisms, thermoelectric applications, X-ray and medical equipment. This exhibition was of particular value in 1962 since it enabled the latest advances in a fast-moving industry to be shown in a way which would not have been appropriate in a commercial trade show. The idea of linking a technical symposium with a relevant exhibition was one which continued to be developed over the next few decades and is now a common occurrence in professional engineering societies.

10.6.1 The Physical Society

Three professional organisations that held yearly exhibitions on their technical interests at a time when the technology was changing rapidly were the Physical Society, the Royal Television Society and the Institute of Patentees.

The exhibitions presented by the Institute of Physics showed new techniques and displays of available technical apparatus, concerned mainly with measurement, to its members and other interested technologists. They were held yearly beginning in 1910.

These exhibitions became a recognised meeting ground for the experimental physicist and electrical engineer, providing an opportunity to see and discuss recently manufactured equipment. A process of segregation began to separate exhibits into broad catagories of research and trade exhibits and contributions by university research and government departments. In the last exhibition before the outbreak of war in 1939 the Post Office Research

Department showed the apparatus used for the London–Birmingham coaxial cable transmission route which had just been completed. This incorporated 19 repeater stations designed to handle an effective bandwidth of 0.4 to 2.2 MHz and capable of carrying 400 telephone conversations at once. The equipment shown included an early application of a precision dynatron oscillator for transmission measurements [19].

No exhibitions were held in the years 1940–1945, and the first (the 30th) to be held after World War II in 1946, was a notable one since it provided the first public opportunity to present some of the great strides that had been made since 1939 in scientific invention. Writing at the time, M.G. Scroggie of *Wireless World* comments,

> '... *to anyone coming fresh from 1939 the atomic bomb itself would have been less unexpected than the Magnetron, weighing a mere pound or so, but capable of generating 500 kW at 3,000 MHz via a single turn coupling coil only 0.64 cm in diameter ...*' [20].

He goes on to comment on other effects of wartime technology: the availability of 'tropical finish' on electronic equipment of all kinds, leading to its adaptability for temperature extremes and rough handling; miniaturisation including very small valves; and, above all, the ease of operation of complex technical equipment where aids to correct operation included the use of a 'magic eye'—a miniature cathode-ray tube providing a fluorescent screen area over a varying size of arc, dependent on a controlling voltage—soon to be an essential part of the tuning procedure of postwar radio receivers.

Equipment for testing and measurement in physics, a dominant theme at the Physical Society Exhibitions, was in 1946 confined almost entirely to the radio and electronic field. For the first time the technical public could become familiar with the calibrated oscilloscope (the Cossor 448), with its controls directly calibrated in time and voltage. A new AVO valve tester was shown, able to test a whole gamut of new varieties of valve types and mountings which had appeared in the preceding decade, a signal generator by Marconi (model TF867), providing an input over the then wide range of 15 kHz to 30 MHz, and accurate valve voltmeters having an extremely high input impedance. An important exhibit, not seen previously outside the laboratory, was the Metropolitan–Vickers electron microscope (type EM2), operating at 50 kV, providing a magnification of up to 50 000 times (a resolving power 50 times greater than the best optical microscope of the day).

In 1948 the emphasis was on research and development on radio techniques applied for noncommunication purposes and the use of radar in other branches of physics. The travelling-wave tube was shown by GEC as a new method of obtaining amplification at extra high frequencies. A small magnetron was shown by the same company capable of providing a peak power of 15 kW at a wavelength of 8 mm, and the magnetic (transformer) amplifier made its first appearance in public. The Medical Research Council exhibited their linear accelerator, using a pulsed power magnetron to obtain electron energies of 20 MeV. For the first time the wartime research work by Philips in Holland made its public appearance as Mullard exhibited the properties of a new ferrite ceramic material, 'Ferroxcube',— destined to make a considerable impact in radio and television development in the following decade and beyond.

In the 1950 Exhibition, held at Imperial College, London, whilst transistors were still in the laboratory stage, the first public demonstration was given by GEC Research Laboratories of a three-stage audio amplifier using their new germanium transistor, achieving a gain of 66 dB. Also—a portent of a new direction in calculation—the de Havilland Company showed their analogue computer, solving a set of 12 simultaneous linear equations, whilst AEI were showing the use of resistor networks for the calculation of electrical field patterns. The magnetic amplifier, shown in experimental form at the previous Physical Society Exhibition, was now exhibited by several firms carrying out a number of practical applications.

The Physical Society exhibitions were essentially national rather than international events. A new precedent was set in 1966 when the Centre National de la Recherche Scientifique was invited to hold a display of their instruments. This was followed the next year by a display of instruments provided by the West German Ministry of Scientific Research, and thereafter this form of invited exhibition was to become a feature of the Physical Society exhibition.

The exhibitions ceased to be held on a regular basis in 1975, and the role of specialised instrument exhibitions passed to the IEA and Electrex organisations (see Chap. 14).

10.6.2 The Institute of Patentees and Inventors

Founded in 1919, the Institute was formed to assist members in obtaining patent rights for their ideas and to provide advice on matters connected with inventions and their exploitation. As a method of obtaining a useful input on the value of the patented article and a means of displaying recent inventions to the technical public the Institute arranged yearly an International Exhibition of Inventions. These were held in the Central Hall Westminster and in other London venues, were open to the public, and often contained exhibits of interest to the engineer.

Generally they were arranged in two main sections. One hall contained trade exhibits, that is patented articles which had already reached the manufacturing stage, and a second hall was set apart for inventions which, although patented, had not yet been made in any quantity, if at all. Electrical devices were shown frequently, at a time when whole sections of industry and the home were being connected to the new local and national generating stations. At the 1928 exhibition an example of early production control was seen in a device manufactured by Fidees Controls Co., which was designed to register the activity of every electrically driven machine in a given production area and thus to determine the idle periods for such equipment. The British Smokeless Co. showed an electric grille to cook both sides of meat at once. An apparently topical requirement—a portable apparatus for electroplating, said to be exceptionally useful for worn plating in motor cars—was also shown. An intriguing item was recorded of *'an automatic device which will answer the telephone when there is no human being present',* but no details were given [21]. The 1929 exhibition coincided with the publication of the first issue of the *Inventor,* the Journal of the Institute.

By 1939 a whole gamut of aspiring electrical inventions was available for public appraisal: an electric kettle with a spiral tube inside permitting it to heat (say) milk as well as water; a portable air-conditioning cabinet; an automatic fire alarm system and an electric cooker which utilised the fact that oil vapour is a better heat conductor than air and by using special saucepans was claimed to require 40% less electricity than a normal cooker. A topical invention which made it impossisble for a light to be switched on without drawing the curtains reflected the wartime conditions under which this Exhibition was held [22].

Later exhibitions were held in a number of different venues. The 1951 Exhibition formed part of the Ideal Home Exhibition of that year. The exhibits included a complete home laundry, with an electric drying cabinet (a combined washing machine and spin-dryer did not make its appearance until the 1960s), a rotary iron, and a converter to enable a standard gramophone record player, rotating at 78 rev/min, to rotate at $33\frac{1}{3}$ rev/min to suit the new long-playing records just becoming available.

The last Inventions Exhibition was held in 1953, again at the Central Hall Westminster [23] and, in addition to a wide range of electrical ideas from inventors, featured a number of industrial companies such as GEC (a portable X-ray unit), EMI (a dielectric heater) and Cossor Ltd (a new oscilloscope).

10.6.3 The Royal Television Society

The Television Society was formed in 1927 (it became the Royal Television Society in 1960) for the purpose of furthering the study and development of the technology among experimenters and private users of television, but soon extended its role to informing technologists within the industry of the latest developments and thus, to a certain extent, affecting the design of television receivers. It afforded a common meeting ground for professionals and amateurs interested in current television research. Its members met for reading of papers on the technical aspects of television, often given by members of research laboratories and designers in industry, and meetings were open to the general public. They were well attended during the postwar era when manufacturers were competing strongly with each other and, during these formative years of the British television industry, they played an important role in bringing the latest ideas to designers of the new receivers then being developed.

A special feature of its annual programme was the presentation of exhibitions, which gave a more technical view of television design than could be obtained from the yearly National Radio Exhibitions at Olympia.

The first Exhibition was held in 1929 at the Engineers Club in London under the auspices of its President, Sir Ambrose Fleming, and was dominated by the ideas of Baird and his rotating disc television receiver, although some cathode-ray oscillographs were also shown by Messrs W. Edwards & Co. and a private member. The early mechanical receivers lent themselves to amateur production and many of the exhibits in the first few exhibitions were provided by members of the Society. Typically an amateur receiver would have a small disc and produce a picture about 8 cm square from experimental transmissions. In the 1930 Exhibition, held at University College London, the venue for the next three yearly exhibitions, working transmitting equipment was set up by Capt. R. Wilson (a member) to provide a vision signal by land-line to all the receivers in the hall. Six rotating disc receivers were demonstrated by members of the Society, who competed for a Challenge Cup presented at the Exhibition for the *'most meritorious exhibit in relation to television'*.

The third Exhibition in 1931 began to exhibit receivers suitable for the commercial market, typically with a drum-scanning device to give a 40-line field picture. An amateur set was shown by a Mr Walker, with which he claimed to have received the first transatlantic television broadcast.

By 1933 the venue for the yearly exhibition was transferred to Imperial College of Science, South Kensington, which had already built a reputation for television research in academia. This was at a time when the public television broadcasts transmitted by the BBC occurred between the hours of 2300 and 2330, the inconvenience of which was commented on by Sir Ambrose Fleming in his opening speech. To facilitate a live demonstration at the exhibition the BBC, in conjunction with Baird Television, arranged to transmit on their short-wave transmitter, operating at a wavelength of 7.75 m, some Baird transmissions during the two days of the exhibition, which attracted over 3000 visitors. This became the first Television Society exhibition to include a substantial commercial presence. Receivers and other equipment were exhibited by Baird, Marconi, General Electric, Edison–Swan and British Thomson Houston. It was the first time that three entirely separate methods of television reception could be seen demonstrated together. A rotating mirror drum was shown by Baird and Marconi, a Nipkow disc by W.L. Wright and an early use of an electrostatically deflected cathode ray tube by Ediswan. Some 30 stands were to be seen in the exhibition hall.

No further exhibitions were arranged by the Society until the 1950s, although the Society was to play an important role in the seminal exhibition on the history of television which took place at the Science Museum in 1937 [24].

In the postwar period the Society was active in arranging meetings on the rapid developments that were taking place in the television field with the cathode-ray tube completely usurping the scanning disc receivers of the 1930s. The first public exhibition was not given until 1953 when, in addition to commercial and members' equipment, the Society's own television transmitter was shown. This was designed for propagation tests at 427 MHz using the standard BBC 405 line video waveform. The most advanced television receiver shown in 1953 was an experimental 41 cm table model employing an aluminised tube operated at 12 kV potential, presented by the GEC Research Laboratories.

By 1958 new developments and experimental receivers were demonstrated by commercial manufacturers and amateur equipment ceased to play a part in these exhibitions. Bush Radio showed an experimental colour television receiver demonstrated with input from a test signal broadcast through a radio link from their Kew laboratories or from a Post Office colour generator located on the Post Office stand at the Exhibition. GEC were also demonstrating their colour television receiver, although it was to be several years before the general public were to receive pictures in colour.

The BBC had a significant presence at the Exhibition in 1959 and demonstrated several of their generators and test equipment used by manufacturers for receiver development. For the first time BREMA (the British Radio Equipment Manufacturers' Association) took part and showed a number of tuner units suitable for the new UHF transmission bands for television signals for both 405 and 625 line standards. Only one amateur exhibit was presented by a member of the Society, a 'home-made' colour television receiver using a Mullard 48 cm experimental shadow mask tube. It may have been the lack of members' contributions which made this the last Television Society public exhibition. By the late 1950s it had become a commercial event not so very different from a small trade fair. The aims of the Society itself had begun to change, and in the 1990s it is fully concerned with the technical presentation and management of television programmes and is no longer concerned with the technical problems of equipment designers which dominated the early days of television.

10.7 References

1 BAKER, E.C.: 'Sir William Preece, Victorian engineer extraordinary' (Hutchinson, London, 1976), p. 317
2 MORAND, P. *in* 'Livre des Expositions Universelles 1851–1989' (L'UCLAD, Paris, 1983), p. 113
3 GUILBERT, C.F.: 'The generators of electricity at the Paris Exhibition of 1900' (G. Naud, Paris, 1902)
4 BLONDEL, A.: 'Oscillographes, nouveaux appareils pour l'etude des oscillations electriques leutes', *Comptes Rendus, Acad. Sci.*, 1897, **16**, p. 502
5 TOUGH, JOHN and O'FLAHERTY: 'Passenger conveyers' (Coleman, London, 1971), p. 20
6 'Automobile Club's Show at Madison Square Gardens, N.Y.', *Electr. World*, 1900, **36** (19), pp. 785–86
7 'Electric automobiles at the Paris Exhibition', *Electr. World*, 1900, **36** (2), pp. 55–59
8 MORGAN, P.F.A.: 'Highlights in the history of telecommunications', *Telecom. J.*, 1986, **53** (3), pp. 138–148
9 'Le Cinéorama', La Nature, Paris, 21 July 1900
10 'Glasgow International Exhibition', *The Electrician*, **47**, 29 May 1901, pp. 168–169
11 ADAMS, E.D.: 'Niagara Power, II' (London, 1927), p. 325
12 PASSER, H.C.: 'The Electrical Manufacturers 1875–1900' (Harvard University Press, Cambridge, Mass., 1953)
13 BLYTHE, S.: 'Buffalo and her Pan-American Exposition', *Cosmopolitan*, **XXIX**, 1900, pp. 507–515
14 'Electricity at the recent Buffalo Exhibition', *Engineering*, 24 Jan 1902, **73**, pp. 109–111
15 GOULD, L.L.: *in* FINDLING, J.E. (Ed.): 'Historical dictionary of world's fairs and expositions, 1851–1988' (Greenwood Press, New York, 1990), pp. 165–171

16 STARR, R.: *in* FINDLING, J.E. (Ed.): 'Historical dictionary of world's fairs and expositions, 1851–1988' (Greenwood Press, New York, 1990), pp. 219–230
17 'International transistor exhibition', *The Engineer*, 1959, **27**, pp. 802–803, 853–855
18 'AIEE Electrical Engineering Exposition', *Electr. Eng.*, 1962, **81** (March), pp. 226–227
19 'The Physical Society Exhibition', *J. Tv. Soc. Ser.* 2, 1939, **2** (Dec.), p. 498
20 SCROGGIE, M.G.: 'The Physical Society Exhibition', *Wireless World*, 1946, **LII** (Feb.), p. 47
21 'The Institute of Patentees Inventions Exhibition for 1928', *Electr. Times*, 18 Oct. 1928, **74**, pp. 553–554
22 'Exhibition of Inventions', *Electr. Rev.*, 1939, **124**, pp. 292–293
23 'Inventions Exhibition', *Electr. Rev.*, 1953, **27** (Feb.), p. 377
24 'Television Exhibition at the Science Museum, London', *J. Tv. Soc. Ser.* 3, 1937, **2** (June), pp. 265–272

Chapter 11
Exhibitions between the Wars

11.1 The 1920s and 1930s

We now need to go back a little and look at the period between the two World Wars which had a special significance in the way technology was presented to the public at this time. At the beginning of this period most countries holding international exhibitions were concerned to show that the recovery from recession was on its way, and at its end that their industry and people were sufficiently recovered to face any preparations for a war that most expected to come.

In the democratic mainland of Europe this was to a large extent expressed for national and international exhibitions in the concept of Empire, to show the efficiency, productivity and humanity of the colonial system, as it turned out, for the last time before these ideas were overcome by war and the resurgence of nationalism. A similar theme was to dominate British exhibitions in the same period, but this was more concerned with the concept of Commonwealth and had a domestic significance, considered in the following Chapter. The totalitarian states in Europe had at this time discovered the value of an international exhibition to make a political statement and made this quite strongly at Düsseldorf, Barcelona and, somewhat surprisingly, at the International Exposition in Paris in 1937.

The United States was perhaps the most idealistic, and considered in its exhibitions the idea of the application of the scientific method as a way out of the Great Depression of this period, and expressed this by several vigorous technological expositions in Philadelphia, Chicago, San Diego, Texas, San Francisco and New York. This was a part of the 'New Deal' policy initiated by President Roosevelt and led to his strong support of the exhibitions, particularly in the 1933

'Century of Power', which was extended into 1934 at his recommendation [1].

11.2 Barcelona, 1929

The initiative for this Exhibition arose in 1913, when a group of electrical managers in Spain proposed an international electrical exhibition to demonstrate the great progress that had occurred in their industry since the 1880s. The city of Barcelona was no stranger to international exhibitions, having mounted its first international exhibition, in which full use was made of electrical lighting, as early as 1888 (see Chap. 8). Realisation of this second international exhibition was deferred for a number of years and, when finally held by General Franco as part of his political resurgence in 1929, became one of the most spectacular events of the interwar years. Its main feature was its lighting, which was superb:

> *'The great line of Art Deco style glass lights leading up to an enormous illuminated fountain with its water cascades just below the main exhibition building on Montjuinch, were never to be surpassed'* [2].

The lighting effects were engineered by the American company, Westinghouse, under its lighting manager, Charles J. Stahl, and represented the first extensive use of mobile coloured lighting, which made the Barcelona exposition so remarkable [3]. The method of producing the multicoloured lighting displays has been illustrated by Stahl through considering what happens over the 600 m of the main avenue of the Exposition, the Avenida de Americas looking towards the floodlit National Palace. This is shown somewhat inadequately at night in Figure 11.1, which lacks the colour which made the lighting effects so spectacular. Numerous concealed floodlights were sited over the entire area, with projectors furnished to provide either blue, red or white light at about 40 m intervals. The colour and intensity of each projector could be separately controlled, with the overall control programmed so that, for example, the area may be first flooded with blue light which gradually gives way to lighter blue as more white light is added, then to pink and finally to red as all other colours are dimmed, the entire change taking place slowly in ~20 min. These changes could also be progressive, starting with blue at the beginning of the avenue and reaching towards red at the other end,

Figure 11.1 The Avenida de Americas looking towards the floodlit National Palace at the Barcelona Exposition of 1929 (Trans. Illum. Eng. Soc., 1929, 24)

whole having the form of waves of colour sweeping down the avenue, again taking several minutes to complete the cycle. Some 4000 floodlight projectors were employed, each containing 1000 W gas-filled lamps.

Glass columns, fountains and many waterfalls (cascades) formed part of the exterior display. These, too, were floodlit and lit with multicoloured searchlight groups and, for the cascades, illuminated from below the water and from behind, using specially designed waterproof units; they were caused to change colour and intensity slowly over a lengthy time cycle whilst the water pressure was adjusted.

All of this required considerable electronic control and was carried out via a number of complex control centres. At the largest of these the following equipment was installed:

> eight 30 kW 115 V DC, 125 V AC air cooled reactors
> 24 28 kW 115 V DC, 125 V AC air cooled reactors
> 24 24 kW 115 V DC, 125 V AC air cooled reactors
> eight 21 kW 115 V DC, 125 V AC air cooled reactors
> eight 16 kW 115 V DC, 125 V AC air cooled reactors
> panelboards for protection of the individual circuits
> one set of oil circuit breakers, disconnecting switches, fuses, etc.

Three further control substations were situated in the grounds, each containing a number of DC air cooled and oil insulated reactors, DC motor generator sets and a number of panel controls for individual circuits. It was found necessary to use reactance dimmers for circuits rated at >5 kW since dimming equipment using resistors would have been prohibitively expensive (thyratron control was to be used at later exhibitions, such as the Chicago 1933 Exposition). Whilst some of these cycles of lighting effects could be programmed to repeat after ~20 min intervals it was also found possible to achieve a much longer programme, in conjunction with variations in the pressure of water supply, to obtain up to 6 h of unrepeated cycles of changes. Thus it became possible for visitors to be confronted with a different lighting and display pattern with each nightly visit, and this added enormously to the public attraction for the Exposition.

These impressive lighting effects (achieved without the aid of a programmable computer in 1929), to some extent diverted interest away from the exhibits to be seen in the various exhibition halls.

The Montjuinch Park siting for the industrial exhibition was impressive. Apart from the large and palatial domed building seen in Figure 11.1, 16 groups of buildings formed the main part of the

Exhibition. Each contributing country had its own pavilion: France, Germany, Britain, Italy, Sweden, Norway, Switzerland, Austria, Denmark, Belgium, Japan and Finland. The United States' contribution was confined to the lighting arrangements for the Exhibition discussed above. Communication with the centre of Barcelona was by means of an extension to the underground electric railway from the central square.

The most significant electrical features at the Exhibition were transport and electrical power generation [4]. Other areas show little of electrical interest that had not been seen at the Wembley Exhibition of 1924, although collectively a large number of exhibits were shown from the neighbouring European countries, with Germany having the strongest representation [5]. In the grounds two electric locomotives were shown, one using Oerlikon electrical equipment from Switzerland and the other fitted with Brown Boveri apparatus, also from Switzerland. The French exhibited an electric goods locomotive by the Midi company. Several steam locomotives were shown.

Another major part of the Exhibition was devoted to hydroelectric power supplies with models, and information on systems in the Spanish, French, Italian and German mountain areas. The Spanish supply system incorporated at that time 14 hydroelectric power stations located in the Pyrenees, and details of the 100 kV transmission systems were shown.

11.3 Antwerp and Liège, 1930

In 1930 Belgium wished to take an opportunity to celebrate its century of independence from Spain in 1830 with an International Exhibition to show the industrial and cultural developments in Belgium over this period. This was carried out by arranging *two* exhibitions, to be carried out simultaneously: one in Antwerp emphasising the colonial and maritime past and the country's art, and one in Liège to show the nation's prowess in science, industry and the economy.

The Exposition at Antwerp proved the more successful, mainly on account of the Flemish art, colonial exhibits and the willingness of foreign countries to exhibit their contributions using similar themes. (The British pavilion, for example, illustrated the current resources of the British Empire and its connection by sea routes, moving model ships over a large illuminated map of the world.)

Liège was concerned to show Belgium's progress in engineering, mainly mechanical engineering and particularly mining and metallurgical industries, with separate pavilions for transport, civil engineering, the gas industry and a Palais de l'Electricité. A large portion of the ground floor space was occupied by the Association des Constructeurs de Matériels Electrique, the Association des Centrales Electriques en Belgique and the Union des Exploitations Electriques [6]. Many of the electrical exhibits related to heavy engineering, examples being an electric travelling crane and a large electrically driven compressor intended for use in a colliery [7]. Very few exhibits were included from other countries, an exception being Italy, which showed a portable electricity sub-station mounted on a railway bogie waggon having two sets of six wheels capable of supporting its weight of 8.5 t. It was designed to operate at alternative frequencies of 16.7 and 42 Hz at 60 kV and a loading of 2000 to 3000 kVA.

11.4 Paris International Expositions of 1931 and 1937

Two quite different exhibitions were presented in Paris in the 1930s. The first of these was the 1931 'International Colonial Exposition'. This was an Exposition devoted to the idea of 'Empire' and presented a similar view to that presented in 1889 and 1900, but this time entirely constrained to a colonial theme to demonstrate the achievement of the French in managing their colonies in Africa and the East. Britain and Germany did not take part to any large extent. Full contributions were made, however, by other colonial powers: Denmark, Belgium, Italy, Holland, Portugal and the United States. Each country had its own national pavilion. A 60 ha site was chosen at Vincennes to the east of Paris.

Apart from the colonial presentations there was a significant industrial presence. In the Palais d'Industrie the latest products of the electrical, metallurgical, mechanical, automotive and chemical industries found a prominent place.

A more complete picture of contemporary technology was seen at the Paris International Exposition of 1937. Originally this was conceived as an 'International Exposition of Arts and Techniques in Modern Life', but with the precarious state of the national economy at the time the arts contribution was much reduced and the theme of 'economic recovery and the struggle against unemployment' was finally adopted. Although to some extent dominated by the political needs of the time, the French paid considerable attention to technology, with a variety of pavilions and palaces, each one devoted to one particular theme. There were complete pavilions dealing with aluminium, its production and use, plastics (a new technology at the time), the production of linoleum, cinema photography, production of gramophone records, transport, and impressive 'history of electricity' and 'palace of discovery' pavilions. Over 53 halls were devoted to sections dealing with all the sciences as well as many branches of technology.

Somewhat larger 'palaces' contained exhibits on radio, with daily demonstrations of television, not as yet a public service in France. This included Henri Chétien's public demonstration of large-screen cinema (90 m × 10 m), using an anamorphic lens, which we today recognise as Cinemascope, and a Palais de l'Air—a vast hall displaying the products of the developing aeronautical industry. In the 'palace of discovery' a large van de Graaf generator was shown, with many physical and technical demonstrations and 'a room of decimalisation', demonstrating the operation of the early digital calculating machines. The van de Graaf generator was a working double generator, capable of producing electrostatic charges of more than 4 MV. It consisted of two towers of insulating material, each supporting a metallic sphere 3 m in diameter. One of the spheres was charged positively and the other negatively. Six belts of rubberised tissue, travelling at high speed, conveyed a charge to the inside of the spheres. The spectators were able to observe (and listen to) a discharge of between 3 and 5 m in length between the spheres, safety to the public being assured by completely enclosing the generator within a large Faraday cage. Another electrostatic generator was exhibited in which a cloud of fine dust was charged by passing it through a network of fine wires, maintained at a potential of 8000 V. The particles give up their charge at an insulated collector, bringing this up to a potential of ~1.5 MV. This particular demonstration was designed to illustrate the principle of electrostatic precipitation of dust and smoke, an application somewhat ahead of its time in the industrial world of the 1930s. Other experiments were shown in the palace concerned with vibrations, which in this range of exhibits could be mechanical, electrical, visual, audible, etc. Amongst them were shown the phenomenon of the thermionic emission and the development of the thermionic valve. The properties of 15 cm Hertzian waves were also demonstrated.

A special continuous current dynamo, designed by Poirson, using the principles of a Foucault apparatus was shown. It consisted of a copper disc revolving between the poles of a magnet with current drawn from brushes situated at the centre and near the periphery of the disc. This was used to provide a low voltage supply of 10 V at a high current loading of up to 50 000 A and used to provide a working demonstration of electrolytic deposition of aluminium or copper on to larger surfaces. It was also used to show Zeeman's phenomenon and the magnetic change in the wavelength of spectral waves as an alternative to the spectroscope. Elsewhere the Exposition contained items relating to the physical sciences, such as the molecular absorption of light, fluorescence, phosphorescence and the progress made in X-rays. The entire Exposition excelled in the educational possibilities of exhibitions and can be compared with the earlier British 1925 Exhibition at Wembley, which had a similar display of popular science [8] (see Chapter 12).

The electrical installation for power and lighting purposes required for such a large technical exhibition was large. Power from the national grid system was supplied to three transformers on the site, each capable of handling a loading of 590 kW at 12 kV. A separate lighting and emergency transformer of 500 kW rating was also installed, together with two batteries of 400 Ah accumulators as an emergency reserve.

Powell, of the Illumination Engineering Society, notes that the installation for the exposition lighting was laid in a hurry, and that

> '... no attempt (was made) at concealing wiring ... and (it was) laid along the surface of the ground, at the edge of pathways, in grooves in the sides of buildings etc. and no attempt made to conceal the floodlighting equipment' [9].

Nevertheless some interesting lighting features were seen for the first time at this Exposition [10]. These included 1000 W trough reflector lights with rotating colour screen drums, mounted in watertight compartments for use beneath the many fountains throughout the grounds. Much use was made of tree illumination along the Quai d'Orsay. Scattered amongst the foliage were distributed 600 projectors, 150 sodium lamps, 225 mercury-vapour lamps and 225 tubular incandescent lamps, a total of 180 kW of lighting power served by 50 branch circuits, each having a unit capacity of 7 kVA. On the Bridge of Alexander III across the Seine a 'triumphant way of light' was contributed by the Philips company. This consisted of 12 unit columns of aluminium supporting high-intensity mercury lights combined with floodlighting. The mercury lights at the centre of each column were each of 500 W providing 20 000 lumens of light. In each unit powerful loudspeakers provided a classical music repertoire. The opportunity was taken to floodlight the Eiffel Tower anew. Seven hundred and fifty incandescent projectors, each of 1000 W, were affixed to the structure using red, yellow and green coloured screens. A total of 8 km of luminous tubing was located in the four base arches, consuming a total of 200 kW. Artistic water displays of very many jets and fountains with both water pressures and colours were controlled from a device looking like an organ console and played as such to delight the crowds. A grand hall of electricity was floodlit to display a gigantic mural 10 m × 60 m showing a picture history of physical and electrical science[1]. Each night an elaborate fête was played out with different co-ordinations of music, colour, fountain control, motion and mechanical effects of released smoke, gas-filled balloons and fireworks from barges moored in the Seine, converting the Exposition into a fun fair. A total of 61 500 kVA of electrical power was needed to sustain the entire Exposition.

Forty-four nations contributed to this Exposition, which attracted 31 million visitors. The largest and tallest stands were those of the Russian and German pavilions, facing each other across the avenue, the latter designed by Albert Speer and dominated by a huge Nazi swastika. It was reported that Speer obtained a glimpse of the Russian plans for the Exposition during a preliminary visit to Paris and arranged the height of the German pavilion accordingly!

11.5 Düsseldorf 'Nation at Work' Exhibition, 1937

In addition to its Pavilion at the Paris Exposition of 1937, the German government presented a large national exhibition of its own at Düsseldorf in the same year. Both had the same general purpose—to present Germany as a leading industrial power in a Europe confused by conflicting signals about its future intentions. The exhibition at Düsseldorf, which opened on 8 May, was designed to show the German 'Nation at Work' in the midst of technical developments that had taken place since the announcement of Hitler's 'Four Year Plan' at the Party Rally in Nuremberg in 1936. It had a strong political message to emphasise the success of a programme of national self-sufficiency with the goal to utilise to the full Germany's own raw materials and to exclude as far as possible all materials of foreign origin [11]. A few foreign pavilions were arranged, including a major one from Italy which displayed every aspect of life under Mussolini,

---

[1]'La Fée Electricité', subsequently moved to the Musée d'Art Moderne de la Ville de Paris

with the aim of promoting tourism. Although predominantly a national exhibition it was also concerned to play an important role in international trade and, partly on this account, attracted over 5 million visitors until it closed its doors on 17 October.

The Exhibition covered not only industry and commerce, but also town planning, housing and garden design. It included a large amusement park, facilities for concerts, and evening entertainment. This latter included an extensive display of fountains, illuminated by ever-changing coloured lights. However, industry was the main theme and it gave overseas visitors a startling insight into the breadth, energy and ingenuity displayed in the production of substitute materials for all purposes together with an increasing use of light alloys and plastics in engineering design.

Düsseldorf had been the site of several exhibitions since 1900, the two largest being the Industrial Exhibition of 1902 and the 'Gesoli' Exhibition of 1926 (both trade fairs). The 1937 Exhibition covered 77 ha and consisted of 42 halls, together with an exhibition town of 96 working class houses and garden plots. A similar desire to display 'workers' houses' had been expressed by Prince Albert at the 1851 Great Exhibition when, as 'President of the Society for Improving the Condition of the Labouring Classes', he caused to be built four 'model houses' at the edge of Hyde Park [12].

The entrance to the exhibition led into the 'hall of honour', which extolled the work of the various engineers, chemists and others who had worked to produce the range of transformations carried out on certain raw materials to increase their industrial value. Examples were demonstrated, such as the conversion of iron into high quality watch springs, coal into chemicals and solid fuels into liquid fuels—a major German accomplishment—and synthetic rubbers of all kinds. Many of the halls were devoted to heavy engineering and generously equipped with working examples of the machinery used. The hall of steel and iron contained working furnaces, rolling mills and other equipment. These were housed in a massive building with an 85 m unsupported roof span and a length of 45 m. Several electrically controlled cranes ran its length, controlled either by the moving cab or, for demonstration purposes (e.g. pouring molten metal), from a stand in the hall. The hall of machinery was arranged to demonstrate modern machine shop practice, and amongst other impressive machines contained a working multiple boring machine capable of handling a workpiece up to 6 m in diameter and weighing up to 50 t. A high-speed vertical boring machine, a planer with a machining length of 10 m, a roller grinding machine and other grinders were demonstrated, all electrically controlled, some with a Ward–Leonard drive to give a wide range of table speed. The demonstrations in this hall concluded with a horizontal forging press busily engaged in manufacturing bicycle hubs at the rate of 100/h.

In the hall of light steel and alloys the metallurgists were exhibiting the special steels which did not contain nickel (a material difficult to obtain within Germany). Instead they used a number of substitutes, such as manganese, and claimed to achieve excellent results with these. The uses of some of the newer light alloys were shown in cable making and in the electrical industries to minimise the use of copper. The hall of synthetic materials continued the theme, with the new synthetic materials recently developed in Germany. More than 200 well recognised substances for industrial use were shown. A new fibrous material with a phenol base was one of the new plastic materials to emerge from this development work. This was shown used for gear wheels and bearings for machine tools. A vital synthetic invention for Germany at that time, that of synthetic rubber known as 'Buna', made its first appearance at this Exhibition, and was suitable for car tyres and other applications. A large hall of 60 m in length housed the railway exhibits. One of the purposes of showing a range of locomotives and rolling stock was to demonstrate the use of synthetic materials and new uses for available materials in the construction of these exhibits. One example shown in detail was a new magnetic safety track brake, actuated when a signal is passed. This hall was one of the stations on 4 km of narrow-gauge passenger railway encircling the grounds.

11.6 'A Century of Progress' Exposition, Chicago, 1933

The Chicago World Fair of 1933 was the forerunner of a group of American exhibitions, promoted in the interwar years, extolling the virtues of the scientific method and the part played by industry in the new scientific and technological development. Succeeding exhibitions were held in San Diego 1935, Dallas 1936–1937, San Francisco 1939 and the New York World Fair in 1939. All emphasised the part played by science in the modern world and the contribution made by the industrial laboratories and workshops throughout America. The enthusiasm generated at the time is clearly seen in an editorial for *Electric Light and Power*, when writing about the 'City of Tomorrow' at the New York World Fair in June 1939,

'... low buildings, exterior murals, colorful decorative effects, spacious gardens, wide boulevards, soft lighting—all promise the happier, more pleasant living that will be the result when our present scientific knowledge can be applied universally to man's quest for leisure, comfort, security and culture' [13].

The year 1933 marked the centenary of the founding of the city of Chicago, and the Exposition was to demonstrate the scientific and industrial progress over that period. The exhibition theme was set by the bizarre method of its opening. Several astronomical observatories around the United States arranged to beam the light from a star, Arcturus, on to photoelectric cells to initiate electric impulses. These were conveyed to the exhibition grounds via telegraph lines and caused to actuate the formal opening of the Exposition, and for each night that the Exposition remained open, to switch on the lights in the grounds. Arcturus was chosen for its distance, 40 light years away, so that symbolically the light commenced its journey at the time of the opening of the 1893 World Columbian Exhibition, also held in Chicago [14].

The organisation of the Exposition was guided by the National Science Research Council, established in 1916 to facilitate co-operation between science, industry and the military, and which had now become influential in co-ordinating scientific research amongst industry and academia. An educational theme was predominant throughout the entire Exposition. It was set up over a narrow strip of land adjoining Lake Michigan and stretched 5 km along its shore, covering an area of 173 ha. A major building was the hall of science—two floors of exhibits devoted to mathematics, physics, chemistry, biology, medicine and industrial applications of science [15]. Some manufacturers and foreign countries had their own pavilions. Many, such as General Motors, Ford Company and Sears Roebuck, were huge. The Ford Motor Company Pavilion was 273 m in length with a rotunda in the centre displaying a 6 m globe, the inside of which indicated Ford's international operations. A fully manned final assembly line for Ford cars was set up inside the building, with galleries where spectators could look down on the progress of assembly.

Exterior lighting on all the buildings was carried out by the combined efforts of Westinghouse and General Electric and consisted of a variety of functionally designed fixtures that provided white light at ground level and coloured light directed on to the walls of the buildings at higher levels. Most of the buildings had some kind of distinctive feature, often a tower or pylon. The Electrical Pavilion was flanked by two 30 m pylons framing a water gate, through which the visitors could enter the pavilion by boat from across a lagoon. In the grounds the balloon gondola of August Picard was shown, in which he had recently ascended to the record height of 14 600 m.

The electrical industry participated in a large diorama, called 'electricity at work'. This was designed to bring to visitors the many and varied tasks that electricity can provide. It took place in the Electrical Building and covered 744 m² of the first floor. The visitor first encountered a historical exhibition, commencing with Hero's turbine, Watt's engine, Volta's experiments, Oersted's devices and Faraday's dynamo. A main theme lay in the use of electricity in the home, in industry, in business and on the farm. Most of these displays consisted of equipment for operation in a full-sized installation, except for the larger exhibits, such as the representation of a complete steam generating plant. The use of electricity in the home was demonstrated in three full-sized displays of a living room, kitchen and basement. This latter included laundry, air-conditioning and other utilitarian equipment, new to the housewife of the 1930s. Further emphasis on domestic exhibits was seen in the 2 ha of lakeside space given over to 13 full-sized model homes in a section entitled 'house and industrial arts'. In one of these vinyl plastic was exhibited for the first time. It was used lavishly to line the walls, floors, furniture and fixtures of a 'house of the future', designed to demonstrate the contribution of electricity and modern technology to the 'American dream' [16].

Farm electrification was demonstrated, including soil heating and irradiation in the greenhouse as well as equipment for milking, dairy operation, heating, ventilation, etc., all of which were to be shown in much more realistic conditions on the half-hectare 'electrical farm' forming part of the 1939–1940 New York World Fair, described below. An outdoor display by the International Harvester Company consisted of a radio-controlled farm tractor seen crossing a field, manoeuvring and returning under the direction of an operator at the edge of the field.

The business displays followed the same format with five model rooms, designated as a butcher's shop, grocery store, bakery shop, restaurant and beauty parlour, all incorporating the lighting and appliances relevant to the particular application. Industrial applications were dominated with the recent discoveries and techniques in electrometallurgy, with electric furnaces for melting, heat treating and annealing metals. Other uses included welding (also demonstrated on refractory metals in the travel and transport building) and the control of power applications through the use of a phototube and relay.

Elsewhere in the Exposition the hall of science contained much on electricity and another historical display. This was also a diorama, having many sections, beginning with Benjamin Franklin with his somewhat risky kite experiment charging a Leyden jar with atmospheric electricity and continuing with Oersted's experiments showing the relationship between electricity and magnetism. The discoveries of Volta, Galvani, Faraday, Henry, Ampere and Edison were shown in some detail, the whole display presenting a short course in electrical technology, expected to be absorbed before the electrical building was entered [17]!

A number of electrical companies provided their own educational features to the Exposition. Bell Telephone Laboratories demonstrated and displayed communication devices, scrambled speech, delayed transmission, graphic transmission, etc. The Edison Power Company and four other companies providing electrical power to the city of Chicago combined to provide a huge pictorial map, translucent and illuminated from below. Moving a narrow pencil of light over this map located the position of power houses and substations, whilst a recorded message provided appropriate information and a side screen showed a picture of the installation. Westinghouse set up a number of research demonstrations which could be operated by the visitor—an idea now common in science museums throughout the world, but fairly new in 1933. Another Westinghouse demonstration was of Nikola Tesla's experiments in the transmission of powerful high-frequency radio waves to effect the illumination of lamps situated some 6 m from the transmitter. Next to this was an X-ray machine which the visitor could operate to obtain a display of *'the bones of his hand or his closed purse'*.

The lighting requirements for the Exposition were somewhat different from earlier exhibitions in that the buildings were not only large but windowless, requiring permanent and extensive illumination both within and outside the building—the latter in terms of floodlighting on the large outside coloured wall surfaces. At least 3000 kW of lighting equipment was used in the main layout of the Exposition, using incandescent lamps, arc searchlights and gaseous tubes, as well as the additional lighting considered necessary by the scores of concessionnaires using the site. More than 15 000 incandescent lamps, ranging in size from 10 to 3000 W, 24 91 cm arc searchlights, 72 61 cm incandescent searchlights and 30 000 m of neon and mercury vapour tubes were used [18]. The method of switching on these building lights using a light signal derived from Arcturus was as spectacular as anything in the Exposition. A searchlight on top of a 200 m tower picked out each major group of buildings in turn in its beam and initiated the lighting display by photocell control until the entire site was ablaze with light. Elsewhere in the grounds other remarkable light effects were created. One installation, referred to as the 'scintillator', located on the shore of Lake Michigan, consisted of 24 1 m-diameter arc searchlights, capable of changing colour using gelatin screens, and angled by a group of operators, following a given lighting time schedule, to produce programmed lighting effects, similar to the Pan–Pacific Exposition of 1915. A total of about 1440 million cp was directed over the exhibition area in this way. Additional effects were provided by multiple jets of water, fountains and, at gala times, fireworks and smoke clouds, towards which the coloured searchlights were directed. Other lighting effects were controlled by banks of experimental thyratrons, then a comparatively new method of current control.

11.7 Texas Centennial Central Exposition, 1936

In 1936 the State of Texas celebrated its formation, 100 years earlier, by holding an exposition in Dallas, the Texas Centennial Central Exposition. It was not a large exhibition and was remarkable mainly for the prominence it gave to electrical illumination.

The Exposition was built on a site of 71 ha, used for the annual State Fair of Texas since 1886, and consequently it already had a nucleus of permanent exhibition buildings, which were added to for this Exposition. This degree of permanence influenced the use of electricity at the site, particularly illumination, in a favourable way, making it popular with the public during the evening. Most of the important state buildings were incorporated within the Exposition site, such as the Hall of the State of Texas, a 'cotton bowl' sports arena, various museums, an aquarium, a hospital station, etc., and all included in a lavish floodlighting plan. The semipermanent interiors and new buildings for the Exposition were divided into halls, corridors and exhibition spaces and provided with effective interior illumination, since many of the buildings, like those in the Chicago Exposition of 1933, were without windows. Much use was made of indirect alcove light, sometimes using recessed parabolic reflectors, and lighted panels and grilles. The entire interior lighting arrangements were extremely complex and showed evidence of careful unhurried design [19], unlike many exhibitions, where the deadline for completion of the electrical infrastructure always ran short of time. Cycloramas, dioramas, photomurals, backlighted pictures and small cinema screens were required by many exhibitors. There were 14 theatres, often with air-conditioning facilities, each seating 30 to 300 people.

The exterior floodlighting was extensive with much use made of coloured lenses of red, amber, green and blue and coloured cement on the outside walls of buildings. For the illumination of exterior statues and walls, superimposed shafts of coloured light were made to change gradually under electronic control to produce slowly changing irridescent patterns, which were most effective. Some 870 kW was used for this purpose out of a total 3250 kW exterior lighting load. Street lighting and coloured floodlighting for decorative fountains required a further 1500 kW, and the refrigerated air conditioning equipment required about 3500 kVA.

Electrical power required for the Exposition was transferred from the local generating station through three 13 000 V supply lines, one of which was carried via a separate route to act as an emergency supply. Two substations on the site each contained two 3-phase transformer units, supplying 4000 V at a capacity of ~4500 kVA. Site distribution was nearly all underground—one advantage of a quasipermanent site. In the hall of electricity, television was demonstrated *'with awe and wonderment"* [20], and live radio broadcasts were made from the Exposition site. General Electric included a 'house of magic' showing a future domestic life with electricity carrying out a multiplicity of house tasks. Western Electric demonstrated its 'electric eye' detector to initiate various processes [21] and a 'talking clock', possibly for the first time. The fledgling IBM Company exhibited their electric writing machines together with accounting and bookkeeping devices and time-keeping instruments. Bell Systems gave an early demonstration of 'speech scrambling' on its telephone network [22].

11.8 San Francisco 'Golden Gate' Exposition, 1939

The period just prior to the Second World War was not an auspicious one for staging an international exhibition, yet the United States managed to open two of them, both extensive and impressive: the San Francisco Golden Gate Exposition in 1939 and the New York World Fair held over two years, 1939–1940.

The San Francisco Exposition of 1939 was planned as a celebration of two recently completed engineering feats in California, both of them bridges: the Golden Gate bridge and the San Francisco–Oakland Bay bridge. This latter passed through Yerby Buena Island in the centre of the bay, part of which was to serve as the location for the 162 ha Exposition site.

Electric lighting had played an important role in exhibitions since the Electrical Exhibition held in Paris in 1881, which made much use of a large accumulation of Ediswan lamps and created a remarkable display. Succeeding exhibitions had enlarged the visual impact of such artificial illumination with ever more elaborate lighting systems. This was the case with the Centennial Exposition in Paris in 1889 for its use of effective interior illumination, followed by the Columbian Exposition in Chicago in 1893 which also made wide use of the then novel incandescent lamp and first introduced its use in searchlights. By the first decade of the 20th century the use of exterior flood-lighting had developed, particularly in colour, and had been used at the 1915 Panama–Pacific Exposition in San Francisco, whilst mobile coloured floodlighting was seen at the Barcelona Exposition of 1929. The 'Century of Progress' Exposition in Chicago in 1933 introduced the extensive use of gaseous conductor tubes and thyratron control. These advances were made possible by the invention in 1901 of the fluorescent lamp by Peter Cooper–Hewitt, an American citizen, and by the invention of the neon tube in 1910 by Georges Claude, which stimulated the introduction of coloured fluorescent lamps a few years later. The Texas Exposition of 1936 made electrical illumination the keynote of the Exposition and featured much ingenuity with interior illumination and complex mobile coloured lighting on exterior buildings, fountains, etc., under automatic electronic control.

The San Francisco Exposition of 1939 presented a sophisticated display of coloured floodlighting on a scale never before experienced and, as with the Texas Exposition, this became one of the major attractions of the Exhibition. More than 2500 fluorescent lamps were used for floodlighting, principally in blue, pink, violet and gold colours. These were used in conjunction with ultraviolet radiation, mercury vapour units, fluorescent paint and 8000 white floodlighting units, all under the control of electronic units [23].

Some 225 kW of streetlighting was also used. Whilst the basic illumination systems were static, employing floodlighting, spot-lighting, groundlighting and waterlighting, much ingenuity was applied to mobile lighting displays involving dynamic changes in colour, moving searchlights and changing fountain displays, controlled by thyratrons. To supply these large installations two generating substations were provided, diverting 16 000 kW of power via four 4000 kVA 3-phase units to 43 transformer vaults located throughout the site [24].

Later exhibitions were to make use of all these forms of interior and exterior lighting as a necessary concomitant of the Exhibition planning and consequently accepted by the public as part of what they would expect to see when visiting an international exhibition.

11.9 New York World Fair, 1939–1940

Built on waste ground, a marshy area and rubbish dump known as 'Flushing Meadows', this became a 512 ha exhibition site of some importance. F. Scott Fitzgerald described the area in his novel, 'The Great Gatsby' (1925) as,

> *'... a valley of ashes—a fantastic farm where ashes grow like wheat into ridges, hills and grotesque gardens ... bounded on one side by a small foul river'.*

Work commenced in 1936, and eventually the area became a permanent exhibition park which hosted the first (1939–1940) New York World Fair, the second in 1964–1965, and various other public events in the intervening years.

The theme of the 1939–1940 Fair was 'building the world of tomorrow', and the futuristic architecture was designed to set the mood for the up-to-the-minute exhibits, illustrating recent progress in science, technology, the public services and particularly engineering. It was subsequently referred to as *'an engineer's Utopia'* [25, 26]. The focal point of the Fair was an architectural display consisting of a 60 m diameter spherical building, with no windows, the perisphere, and a 213 m high triangular obelisk, known as the trylon (Figure 11.2). Both were skilfully floodlit, with the latter housing the world's highest light.

The Fair consisted of seven exhibition zones:

1. communications and business systems
2. production and distribution
3. community interests
4. food
5. government
6. transportation
7. amusement area

and two exhibition categories:

Figure 11.2   The trylon and perisphere at the New York 1939/40 World Fair
(Official Guide, New York, 1939)

8. medicine and public health
9. science and education.

As in the San Francisco 'Golden Gate' Exposition, earlier in the year, lighting played a significant part in the exhibition. Whilst many of the buildings, including the perisphere, were windowless, natural lighting was obtained by the use of glass blocks and new translucent building materials, and at night the buildings became '... masses of soft glowing colour'. The perisphere itself (weight 2000 t!) was supported by six steel columns continuously hidden by vast fountains of water. These were illuminated by elaborate lighting arrangements which consisted of newly developed 400 W mercury capillary lamps, in groups of nine, at the base of each column and beneath the water. Continual changes in filter colour for these lamps and the accompanying eight 5000 W flood lamps under electronic control made for a spectacular, ever changing display. This was typical of the considered design effort made by illumination engineers at this time. The outside walls of the windowless buildings were coated with luminous paints which, when illuminated with ultraviolet lighting, again created a spectacular effect.

A new type of floodlighting was seen, consisting of a battery of 16 searchlights, each equipped with 1000 W water-cooled mercury-vapour capillary lamps producing '...*by far the brightest light source yet developed*', and each requiring a gallon of water per minute to avoid overheating. This produced an effect which we now obtain more simply by a laser beam at a tiny fraction of the power then needed.

A new lighting feature not seen at previous American exhibitions was a performance of a 'son et lumière' at various locations. One of these, surprisingly enough, was an indoor display accompanying a large diorama, 'city of light', seen in the Edison building. This presented a coloured, lighted and animated architectural model of New York City and billed as the world's largest diorama [27, 28]. This was a substantial exhibit, 'the length of a city block', containing more than 4000 buildings all showing much detail concerning their contents. The model showed a dynamic function with motion, light, sound effects and music synchronised to a recorded narrative of a complete day in the life of the city. This technique of a recorded narrative for the benefit of the visiting public was first used at the St. Louis Exposition in 1915, but here it was viewed by the visitors situated in a train of 600 chairs, each equipped with an individual loudspeaker and traversing slowly along the exhibit (Figure 11.3). The scale model included six-car subway trains travelling along and through the city streets, motor traffic over bridges, miniature working lifts in buildings, ocean liners entering port, small stage productions and a working hospital operating theatre.

The role played by electricity dominated the Exhibition, with various buildings housing specific electrical exhibits, often confined to one particular American manufacturer. Some of these were:

electric utilities
electrical products
power—electricity and steam building
the electrified farm
communication building
A T & T building
RCA building
Westinghouse Electric building
General Electric building
Consolidated Edison building
hall of industrial sciences
hall of electrical power
hall of electrical living

Figure 11.3   The diorama at the New York 1939/40 World Fair
(Reproduced with permission from *World's Fair Mag.*, 1982, **2**)

and various pavilions where foreign exhibitors included electrical items. To provide power for the site two substations provided 65 000 kVA—enough power for a city of 300 000 [29].

General Electric (GE) presented several 'educational displays'. A spectacular demonstration was that of a lightning discharge of 32 000 A between two spheres when the potential difference reached 10 000 V. This made use of six vertical stacks of capacitors, piled 10 m high. A second discharge display was obtained through a 1 MV 3-phase sustained arc, each leg of a different colour. GE also borrowed from a New York museum a 2800 year old Egyptian mummy which was subjected to X-rays, with the results displayed on a movable fluorescent screen traversing the body.

The American Telephone and Telegraph company (AT & T) displayed a voder (voice operator demonstrator), later to be named a vocoder, which demonstrated realistic synthetic English speech using 50 phonemes. Also, against a huge map of the United States showing the extent of telephone exchanges, then controlling more than 20 million subscribers, the visitor was invited to make a free telephone call from a battery of telephones to anywhere within the United States. Remington Rand showed its punched card tabulating machines, accounting and adding devices, anticipating the advent of the computer in the following decade.

RCA demonstrated facsimile transmission which was hailed as, 'the newspaper of tomorrow'. Also, for the first time in public, television broadcasts were transmitted from the top of the Empire State Building, commencing with a transmission by Franklin D. Roosevelt at the opening ceremony (he was the first American president to make a televised speech) [26]. The transmissions were received in a demonstration room containing a number of console receivers incorporating electrostatic cathode-ray tubes, mounted vertically with mirror lid display. (Unlike in the United Kingdom, television transmission was an ongoing activity throughout the War years.)

A number of Theme Tours made their appearance at this Exhibition, in each of which electricity played a central role. The 'City of Light' diorama has already been mentioned. Others include a series of streets, houses and shops, etc. contrasting the uses of electricity in 1892 to 1939, and an extensive display of electricity in use on the farm. This was spread over half a hectare and demonstrated milking, pasteurising, bottling, cooling, a hen battery, egg cleaner and grader, potato processing, lighting and ventilation. In the farm kitchen various electrical utilities were shown, e.g. range, sink, dishwasher, garbage disposal, coffee making, a toaster, etc. In another 'kitchen' display, mounted by General Electric and named by them 'the magic kitchen', all the labour-saving appliances were seen to come to life and operate, apparently of their own volition, carrying out various tasks.

A major theme tour was a second diorama presented by General Motors, called 'The Futurama'. This was a three-dimensional model of industry, towns and villages in, '... America 20 years from now', occupying some 3400 m^2 at various levels. To see all of this the visitors were conveyed in a specially designed and constructed 'chairveyor', a continuous chain of moving chairs having a total length of 478 m. In this conveyor the track carrying the chairs was mounted on an elevated structure which contained many curves and height changes [30]. It was driven by 23 2 hp motors situated along the track. Each double chair was enclosed in a sound-proofed cabin with a window facing the exhibit. The sound commentary was more elaborate than with the Edison exhibit and related to the actual view seen by the spectator at different points around the exhibit. Consequently it was necessary to arrange some 150 *different* descriptive talks to be fed to each cabin. This was accomplished by specially designed multiple playback equipment consisting of a machined steel drum, 3.5 m high, revolving in the vertical plane and carrying on its periphery strips of sound film, intercepting narrow beams of light. The modulated light beams were transformed into varying electrical currents and fed to loudspeakers in the chair cabins in the correct sequence. The Futurama was a great success and carried some 9.6 m visitors during the course of the Fair. A second, improved, 'Futurama II', was installed at the 1964–1965 New York World Fair, to be described in Chap. 13.

The New York World Fair 1939–1940 opened, for the first year, in April 1939 and closed at the end of October. It reopened in May 1940 and finally closed its doors at the end of October of that year. It attracted fewer visitors in the second year, largely because of the withdrawal of a number of the 1939 participants as a consequence of the War. The Soviet Union withdrew its support, dismantled the Soviet Pavilion, which was a substantial one, and shipped it back to Russia. Over the two years the exhibition had attracted 45 million visitors and, although generally acclaimed a success, too much had been spent on lavish presentation and in the end it was declared financially bankrupt. A major problem had been lack of support received from foreign exhibitors at a time when much of the world was engaged in warfare. The organisation and exhibits, however, remained valuable pointers to the preparation of future fairs in the United States and elsewhere.

11.10 References

1 RYDELL, R.: 'The fan dance of science: American World's Fairs in the Great Depression', *ISIS*, 1985, **76** (Dec.), pp. 525–542
2 ALLWOOD, J.: 'The great expositions' (Macmillan, London, 1978), Chap. 9
3 STAHL, C.J.: 'The coloured floodlighting of the International Exposition at Barcelona, Spain', *Illum. Eng. Soc. Trans.*, 1929, **24** (9), pp. 876–889
4 'The Barcelona Exhibition and British Trade', *Engineering*, August 1929, **128** (3317), pp. 162–163, 324
5 'Catalogo official de Barcelona Exposition'. Barcelona, 1929
6 'The Liège International Exhibition', *Electr. Rev.*, 11 April 1930, **106**, p. 684
7 'The Liège Exhibition', *Engineering*, 1930, **130**, pp. 68, 515–518
8 'Palace of Discovery at the Paris Exhibition', *The Engineer*, 1937, **164** (4269), p. 502
9 POWELL, A.L.: 'Illumination at Paris 1937', *Illum. Eng. Soc. Trans.*, 1938, **33** (6), pp. 566–587

10 KALFA, L.C.: 'Illumination at the International Exhibition, Paris, 1937', *Philips Tech. Rev.*, 1937, **2**, pp. 361–369

11 'Nation at Work at Düsseldorf', *The Engineer*, 1937, **164** (4260, 4267), pp. 425–426, 453–454

12 ERIC DE MARE: 'London 1851—The Year of the Great Exhibition' (The Folio Society, London, 1972), p. 81

13 BAKER, R.K.: 'New York's World Fair 1939', *Electr. Light Power*, 1939, **17** (6), pp. 54–74

14 LOHR, LENOX, R.: 'Fair management' (Cueneo Press, Chicago, 1952), p. 197

15 'Official handbook of exhibits in the division of the basic sciences; Hall of Science (Chicago: A Century of Progress, 1934)'. Chicago, 1934, pp. 9–11, 17–19

16 BUCHANAN, A.E.: 'Synthetic houses', *Sci. Am.*, Oct. 1933, p. 149

17 'Electronics at Chicago's World Fair', *Electronics*, 1933, **6** (6), pp. 148–150

18 RYAN, W.D'ARCY: 'Lighting an exposition', *Electr. World*, 1933, **101** (21), pp. 687–698

19 FIES, J.: 'Electrical features of the Texas Centennial Central Exposition', *Electr. Eng.*, 1936, **55** (10), pp. 1060–1074

20 *Dallas Morning News*, 30 July 1936

21 RYSDALE, K.B.: 'The year America discovered Texas—Texas Centennial '36' (Texas A & M Press, College Station, 1987)

22 'Official Guide Book—Texas Centennial Exposition 1936'. Texas, 1936

23 DICKENSON, A.F.: 'Colour, light and structure at the Golden Gate Exposition', *Electr. Eng.*, 1939, **58**

24 BEAR, W.P., BOKKELEN, W.R., and SNOWDEN, W.: 'Electricity for Treasure Island', *Electr. World*, 1939, **111**, pp. 149–151

25 'New York World's Fair', Official Guide. New York, 1939

26 HARRISON, H.A.: 'Dawn of a New Day: New York World's Fair 1939–40' (New York University Press, 1980), p. 48

27 'Electric Utilities tell their story at New York World's Fair', *Electr. World*, **111**, May 1939, pp. 149–151

28 WURTZ, R.: 'The New York World's Fair 1939/40' (Dover Publications, New York, 1977)

29 SAWYER, W.H., and BROOKS, J.A.: 'Sure power with innovations—N.Y. World's Fair 1939', *Electr. World*, 1939, **110** (13), pp. 41–44

30 DUNLOP, J., and WHITE, W.T.: 'An armchair spectator conveyor guide', *Electr. Eng.*, 1939, **58** (12), pp. 509–514

Chapter 12
Electricity and the public

12.1 The domestic users of electricity

From early in the 1920s, as the expansion of generating capacity took place in the developed countries, the proportion of electricity consumed by domestic users increased. This was very marked in the United Kingdom, after the Electricity (Supply) Act of 1926 had begun the progress of rationalising the 600 separate electrical supply undertakings existing at that time, through the establishment of a Central Electricity Board. This culminated ten years later in the construction of the National Electricity Grid whilst, in the intervening period, domestic usage rose from 271 GWh in 1920 to more than 2000 GWh in the early 1930s [1]. The public interest in the availability, understanding and correct usage of domestic electrical appliances had correspondingly increased and was reflected in the content of public exhibitions devoted to the presentation of technical accomplishment and in other related events.

Some of these exhibitions were concerned entirely with one particular domestic application, such as the series of British National Radio Exhibitions. Others were broader in their concept to include other domestic applications, such as the National Electrical and Radio Shows in New York which, in the 1930s and 1940s, provided comprehensive exhibitions of electrical heating, refrigeration, air conditioning, cooking and laundry equipment, as well as radio and television receivers. A common feature was to include a special exhibit entitled *'home of the future'* —a one-room apartment demonstrating a complete range of domestic utilities. This feature became also a major theme in the British series of Ideal Home Exhibitions, which came to become a yearly event at Earls Court and Olympia exhibition halls.

The importance of domestic applications of electricity had already been seen in the electrical sections of several large international exhibitions. The Chicago Exhibition of 1893 contained a section devoted to 'electricity in the home'. The Paris 1900 and Pan-American Exposition at Buffalo of 1901 both contained major groups of exhibits of domestic interest.

In the period up to 1939 a large number of small local exhibitions covering domestic appliances were held in London and the provinces. We have already discussed one of the first of these in Chap. 9, held in St Pancras, London. Their numbers increased in the next few decades. Hampstead held a large one in 1910 which included not only domestic appliances, such as cookers, vacuum cleaners and electric flat irons, but also wiring systems for the home, electric baths, radiators and even an electric piano [2]. Guildford was one of the many locations holding small electric exhibitions in the 1920s and 1930s [3], and the Electrical Association for Women were to hold many of these in various parts of the country, as we discuss later. Abroad, reflecting the increase in generating capacity in Europe during the interwar years, France provided a detailed look at the domestic electrical appliance market in their 14th International Fair in Marseille in 1938.

Other public exhibitions brought technology and the application of electricity directly to the public in terms of national achievement, such as in the various 'Empire' exhibitions, the Chicago 'Century of Progress' Exhibition in 1933, considered in the preceding Chapter, and most significantly in the Festival of Britain held in London in 1951.

12.2 The Empire Exhibitions

The 'Empire' national exhibitions mounted in Great Britain commenced with the Colonial and Indian Exhibition in 1886 (Chap. 8), and continued with a number of smaller exhibitions on the same theme in 1895, 1899, 1908 and 1911 (which also celebrated the coronation of King George V), culminating in the Wembley Exhibition of 1925. There was also a provincial exhibition, 'The Empire Exhibition', held in Glasgow in 1938, and overseas an 'Empire International Exhibition' took place in Johannesburg in 1936. The Wembley and Glasgow exhibitions were guided by the Empire Marketing Board (1926–1933), set up to foster trade within the British Empire.

12.3 British Empire Exhibition, Wembley, 1924–1925

The Exhibition at Wembley had its roots in the British Empire League proposals in 1902, which determined that not only would the Exhibition display the culture and industry of the countries of the Empire , but it would be

> '*a Family Party of the British Empire —its first since the Great War, when the whole world opened astonished eyes to see that an Empire with a hundred languages and races had but one soul and mind, and could ... concentrate ... all its power for a common purpose*'.

Following a lapse consequent on the First World War, the Empire League proposals were adopted for construction in 1919. A site of 87.4 ha was purchased at Wembley and a permanent exhibition was designed (which continues to be used today for a variety of cultural and entertainment events). The method of construction for the site and buildings is notable as the first major use of ferroconcrete in Britain since the technique was transferred from the United States a few years earlier. The buildings were contained in large triangular plots, with each territory of the Empire given its own area. Some common buildings were built, including a large stadium and three palaces of industries, engineering and the arts. The site was particularly well chosen for external communication. It was served by no less than three railways, the former Great Central Railway, the Metropolitan Railway and the London North-West Railway, as well as the Bakerloo underground railway.

The transport problem for visitors travelling the 15 miles of roadways within the site was solved in two ways: by the use of a series of 'Railodock' electric buses and a new variable speed rail system called the 'Neverstop Railway', designed by two English engineers, B.R. Atkins and W.Y. Lewis [4]. This latter consisted of 88 rail cars, spaced out along the track, each seating 18 passengers. The cars were driven by a spiral shaft acting on rubber guide wheels, the mechanism of which can be seen in Figure 12.1. It was arranged to travel at a maximum speed of 26.4 km/h, slowing down at the turning points and the stations to 3.2 km/h, to allow passengers to board the train, thereafter speeding up until the next station was reached. The variation in speed was arranged by altering the pitch of the spiral, seen in the diagram. The system was powered by 14 electric motors, requiring a total power of ~180 kW. It carried 2 million passengers during the two seasons it was operative and was considered to be so efficient that it ran for the second season free of charge.

Figure 12.1   *The 'never-stop railway' at the British Empire Exhibition at Wembley, 1925*
(Reproduced with permission from *Scientific American*, 1924)

Although a major theme of the exhibition was to describe the natural resources and culture found in all parts of the British Empire it also contained a substantial trade fair and educational function[1]. The exhibition was extremely popular with the almost 18 million visitors recorded. Most of these visitors were, however, attracted by large and spectacular events which took place from time to time during the course of the exhibition, such as a 'pageant of empire', which included 15 000 participants, a searchlight tattoo and several military bands.

It was, however, in the large palaces of industry and engineering that the technical progress of the postwar years was seen most clearly. It offered more than a great collection of exhibits illustrating the advancements made by the leading industries of the time; it also set out to bring to the public the inventions of engineers upon which this progress depended. The names of some of the greatest of the early pioneers were emblazoned over the main entrances to the palace of engineering. The principal entrance to the east was named the Watt Gate. Other gates were: to the south, the Stephenson Gate; to the south-west, the Naysmith Gate, the Arkwright Gate, the Bessemer Gate and the Faraday Gate; and finally, to the west, the Kelvin Gate. The building itself was large, 304 m long and 228 m broad at its widest point, and represented an early triumph of design using only ferroconcrete and steel.

The exhibits were contained in two main divisions: shipbuilding, marine, mechanical and civil engineering in the southern section and electrical and allied engineering to the north. A balanced collection of artifacts was assured in the co-operation between the British Engineering Association and the British Electrical and Allied Manufacturers Association (BEAMA), who arranged the selection and display. Care was taken to show the broad progress made by the leading British industries since the war, and their selection provided a portent of the type of contents to be expected at the annual British Industries Fair, which was to dominate the presentation of British Engineering up to the late 1950s (see Chap. 14). The leading role played by BEAMA in the Wembley Exhibition and the presence of so many electrical engineers at the event prompted D.N. Dunlop, the first director of BEAMA, to establish a World Power Conference as an international forum for discussing technical and regulatory matters in

---

[1]The total population of the Empire in 1920 was 450 million people over an area of 3500 million ha. Its imports to the UK were ~£502 million, and exports from the UK to various parts of the Empire were ~£560 million, so that its trade function in this extended common market was important

the energy field, the inaugural meeting of which was held at the Exhibition.

The exhibits in the Electrical and Allied Engineering section proved most popular with the public and were described in the official catalogue as

'*...a liberal education in electrical engineering and giving an idea of the wonderful ramifications in this industry to walk through this section*' [5].

The scale of the exhibits was set at the entrance, with the public viewing from a gallery a completely equipped boiler house, power house and substation of 6000 hp from the arrival of coal in trucks, automatic stoking of the boilers, functioning of three steam-turbines driving generators, to the switchboards linked to substations situated in various parts of the grounds.

Broadcasting was of much public interest at this time (the British Broadcasting Corporation did not start regular broadcasting until 1922 and the National Radio Exhibitions at Olympia were still a year away), so that the demonstrations of broadcasting given at the exhibition were well attended, as were the stands of leading makers of domestic wireless receivers. These included two-, three- and four-valve receivers and 'crystal' sets (Figure 12.2). A historic demonstration of the power of broadcasting was to occur at the Exhibition. On the opening on St George's Day, the King broadcast through the BBC directly from the Exhibition hall to the nation—the first broadcast by a ruling monarch in Great Britain and the first major outside broadcast [6].

Long-distance communication was still dominated by telegraphy and the Morse code. Marine telegraphy was well developed by 1924, with Marconi exhibiting their range of ship's wireless equipment, including quenched spark transmitting apparatus as well as the more recently developed continuous-wave equipment. This latter was shown installed in three ship's cabins with 1.5, 0.5 and 0.25 kW transmitters. Shore apparatus was demonstrated in a beam telegraphy transmitter. This sends out on a short-wave band a different Morse code letter for every two points of the compass as it revolves. These can be recognised by a suitable receiver carried on board ship enabling an exact bearing to be determined. A beam transmitter of this design had just been erected at Inchkeith, an island in the Firth of Forth [7].

A range of electrical products by GEC included equipment for the generating, transmission, distribution and application of electricity (a national electricity grid distribution system was not established until

Figure 12.2   A Brown advertisement at the 1925 British Empire Exhibition
(Official Report of the 1925 Exhibition, Wembley, 1926)

1930, so that the need for private generating systems was still felt).
English Electric displayed an entirely new conversion machine, which
they called 'a transverter'. This converts a 3-phase 3300 V 50 Hz AC

supply into direct current at 100 kV with an output of 1000 kW. Other rotary converters and switchgear were shown. This company also exhibited a 3 hp light aeroplane, appropriately called the Wren. It caused a sensation a few years earlier at the 1923 trials at Lympne, with its remarkably low fuel consumption of 31 km/litre.

Other companies displayed a wide range of electrical plant, including: BTH Ltd., which at that time had large works premises in Rugby, Birmingham, Coventry, Chesterfield and London (Willesden); Brush Ltd, with a demonstration of the construction and working of its 5000 kW turbogenerator; Vickers Ltd, with an electric furnace for the heat treatment of steel and an X-ray spectrograph; and Eastern Telegraph, at that time foremost in the apparatus used in laying and repairing of submarine cables, using their own fleet of cable ships. A wide range of telegraph apparatus was shown: a keyboard perforator and Wheatstone transmitter; receiving apparatus punching a replica of the transmitted tape; automatic printers; duplex apparatus; and William Thomson's original siphon recorder. Domestic electrical appliances, arranged by the British Electrical Development Association, took the form of a 'model electric house' containing all the heating, lighting, cooking and kitchen equipment then available on the British market [8].

An educational feature within the palace of engineering was the exhibition, and demonstrations provided by the National Physical Laboratories [9]. These were fairly specialised for the general public but served to demonstrate Britain's lead in scientific attainments at this time. They included a demonstration of Lissajous figures on one of the new gas-filled cathode-ray oscilloscopes then becoming available commercially, and an application of the thermionic valve in an electronic micrometer, based on changes in oscillator frequency by slight alterations in the position of capacitor plates; the equipment was capable of measurements down to one-millionth of an inch (0.025 μm). Additionally, the Science Museum, Royal Institute and King's College co-operated to bring together a splendid display of historical and original electrical apparatus, including Faraday's ring, Wheatstone's Bridge, Henry's induction coils, Hughes' microphone, Bell's telephone and Blondell's double oscillograph.

12.4 Johannesburg Empire Exhibition, 1936

This was proposed by the Federation of British Industries to celebrate the golden jubilee of the establishment of the city of Johannesburg. It was essentially an industrial exhibition, consisting of a number of halls situated over a 40.5 ha site, now known as Milner Park. These included a hall of industry, hall of light machinery and manufacturing, hall of heavy machinery, a tower of light, hall of South African industries, a number of pavilions containing contributions from neighbouring states, e.g. Rhodesia and other parts of the Empire, and finally a number of pavilions leased to industry, including that arranged by BEAMA, which housed most of the electrical exhibits from Britain.

The contribution made by machinery exhibits, particularly in mining equipment and steel production, was extensive, but the secondary industries, such as the electrical industry, were well represented. The British pavilion was foremost in this, where much use was made of neon lamps operating at low voltage and high frequency. These were used in a ceiling representation of the Northern Hemisphere and in a huge map of the world studded with lights, both displays acting as dioramas to give the effect of change in the visible sky and movement of shipping and regular air flights linking the British Empire. Altogether some 2500 lamps were used in the British pavilion. The total power requirements of 5000 kW were provided by the Victoria Falls Power Company for the entire Exhibition.

A striking contribution was a 46 m 'tower of light' which housed a naval pattern searchlight of 14 million cp, forming a powerful beam of light visible over the entire city. Within the British Pavilion the major electrical companies of the day were represented. South Africa was at that time the biggest overseas customer for GEC. A 12 m model of a complete power station was shown, previously seen in 1931 at the Faraday Centenary Exhibition in South Kensington Museum in London. This was particularly appropriate for the Exhibition since it was modelled on the South African Klip power station in the Vereeniging district for the Victoria Falls and Transvaal Power Company. The model was one of the many instructional models and dioramas shown throughout the Exhibition, which had a strong educational as well as commercial basis.

At a time when the first television transmissions had just commenced at Alexandra Palace it is interesting to note the prominence given to television at the Johannesburg exhibition [10]. A special television installation, consisting of a transmitter broadcasting to identical standards as those used at Alexandra Palace, broadcast to five receivers incorporating 30 cm diameter cathode-ray tubes set up at different locations in the grounds. Cinema film and slides were transmitted throughout the day, this being the first opportunity for the public in South Africa to witness this new development. An experimental television service was shown in South Africa in 1950, when the first live public television demonstration was given at the Rand Agricultural Exhibition in that year. There was a long wait, however, until a regular public service was established in 1975 with a colour transmission using the European PAL system.

12.5 Glasgow Empire Exhibition, 1938

This was opened by the King in Bellahouston Park, some 3.5 km from the city centre, in May 1938. The park covered more than 70 ha, slightly less than the 1925 Wembley Exhibition, and was filled with pavilions, allocated to the 100 or more exhibitors. The exhibition was formed to provide an opportunity for manufacturers in Scotland to find new markets in the developing countries and the Empire. It was rather more than a trade exhibition, however, since the stands of the dominions and colonies showed a number of cultural exhibits, and an entertainment area was created which included a large and permanent concert hall having 2000 seats. A large pavilion was allocated to British exhibits, two to Scotland, and one to each of the leading Dominions, Malaysia and the West Indies. Fifty pavilions were filled by private industrial firms. Other pavilions were taken by palaces of arts, industry and engineering.

The time of the exhibition coincided with a massive refurbishment of the Glasgow electric tramway system, and over 100 new vehicles were available on the opening day. These had been designed the previous year and in consequence were dubbed the 'Coronation' trams, and provided adequate transport to and from the exhibition site. Lighting was lavishly provided through a 20 000 V underground feeder cable from the nearby Dalmarnoch Power Station. A major substation was created on the site having a capacity of 15 000 kW and serving the buildings via 16 800 kW transformers at 440 V. Some 1500 floodlights were used, 400 ground standards and thousands of other lamps, including neon and discharge lamps, varying from 15 to 1000 W each. Three thousand two hundred low-power fairy lights illuminated the trees, and 400 ornamental lamp standards served to light up the roads and avenues. A feature was made of water and submarine lighting, similar to that seen in Barcelona in 1929, but most striking was a 91 m tower surrounded by hundreds of coloured floodlights, many of them the newly marketed 400 W Osira coloured discharge lamps manufactured by GEC. Owing to its elevated position in the park, this could be seen at a distance of ~160 km (Figure 12.3) [11, 12].

The Palace of Engineering, one of the major exhibition buildings, was *'about the same area as Buckingham Palace'* and lit at night with 143 floodlights—a loading of 63 kW. After the Exhibition closed it was

Figure 12.3   Glasgow Empire Exhibition 1938
(*Light & Lighting*, 1938, **31**)

dismantled and re-erected at Prestwick Airport, where it served as an aeroplane hangar for a number of years. British electrical exhibits made their home in this Exhibition building, dominated by the three major firms: BTH, English Electric and GEC.

BTH showed their range of rectifiers, switchgear and other industrial and domestic equipment. Amongst the industrial demonstrations were shown a 400 A mercury-arc rectifier and a current overload protection device making use of air-blast protection. At a time of peak interest in the cinema a number of BTH cinema projection arc systems were demonstrated incorporating a three-arm mercury arc rectifier to provide a 150 A DC supply. Another rectifier for industrial use was shown, controlled by thyratrons.

English Electric showed a diorama of its five power stations operated by the Water Power Company scheme in Scotland. A steam turbine was also illustrated and examples given of its form of construction. A range of domestic items included cookers, electric fires, water heaters, etc. and, in its industrial section, flame-proof motors of 27.5 hp capacity for use in mines.

In addition to floodlighting all the buildings, GEC showed its latest developments in the industrial field. These included a new 170 hp traction motor supplied for the London Underground and its contribution to the five new hydroelectric power stations of the Water Power Company scheme.

12.6 Marseille International Exposition, 1938

The Marseille International Exposition of 1938 was the 14th of a series of French regional annual exhibitions held to demonstrate the technological developments of France and its neighbours. On this occasion its emphasis was directed towards domestic appliances and wireless which, although not providing a large display, did give a very detailed summary of the technical state of this section of the industry in France at that time. In 1946 after the war Britain was to take up a similar theme in its 'Britain Can Make It' and 'Enterprise Scotland' exhibitions. The Marseille Exposition was one of the last European exhibitions to be held before the outbreak of war curtailed such activities. The international participation was somewhat constrained, with the only significant contributions outside France being from Germany, Belgium, Holland, Sweden, Switzerland and Italy. The Exposition took place in the Palace of the Halls in Marseille, used for a number of similar events. The 110 exhibitors taking part arranged their contributions in accordance with a logical classification indicative of working principle rather than purpose [13]. They were:

1. lamps and lighting
2. apparatus utilising resistance
3. apparatus utilising motors
4. apparatus utilising resistance and motors
5. wireless receivers.

A range of incandescent lamps from various manufacturers was shown together with mercury vapour lamps and decorative lamps for the home. The efficiencies of the tungsten filament incandescent lamps of different power ratings and conditions of use were compared in the form of a table and reached little more than 15% for the designs exhibited.

Appliances utilising resistance meant in practice heating elements and cookers. A feature was made of a totally enclosed 'radiant' heating element, using nickel–chrome wire (Figure 12.4), a combination which had been shown for the first time at the Ideal Home Exhibition of 1930 [14]. Cooking ovens were shown by Calor, Backer, AEG, Alsthon and Siemens. These had a heavy cast-iron construction, not dissimilar to the gas ovens then on the market but with the back burner replaced by an electric heating element. They mostly contained a grille heater and one to four radiant cooking elements. Charcoal and gas cookers were shown elsewhere in the Exposition and, surprisingly, a number of electric cookers in which a *'firegrate of charcoal'* is added which is provided *'for cooking and heating in the winter'*. These mixed fuel cookers included several gas–electricity models, possibly because of the uncertainty in the current retail prices for the two fuels.

Other 'resistance appliances' shown were elements for electric kettles, and larger immersion heaters for hot water and bath water systems. Mention was made of the use of storage heaters making use of 'off peak' services (*utilization le courant de nuit*). Thermostatic control of electric radiators was shown and also of small electric plate warmers.

Equipment requiring motors included fans, refrigerators, compressors, washing machines, floor polishers, etc. Most of the manufacturers were French, with a few German exhibitors. An oscillating electric fan was shown. This was new to Europe, although it had been available in the United States for some time. Washing machines were shown using a single-phase supply for domestic use and larger 3-phase machines for industrial use.

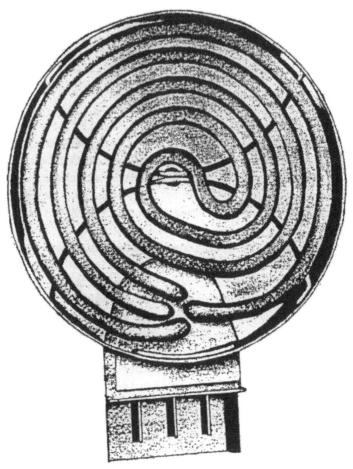

Figure 12.4   A 'radiant' heating element shown at the Marseille International Exposition of 1938
(*L'Electricien*, 1939, **55**)

Appliances shown which incorporated both resistive elements and motors included fan-heaters, hair-dryers and rotary ironing machines. These latter were fairly large devices standing ~$1\frac{1}{2}$ m high with a roller of 60 cm in length and capable of ironing flat sheets at a rate of almost 3 m/min. They would be more suitable for an hotel environment than for domestic use. The market for wireless receivers in 1938 was a rapidly growing one, with some 5 million sets already operating in France. Although French television transmission had started experimentally in Paris, no television equipment was shown at Marseille. A very large number of manufacturers of wireless receivers and components exhibited their products, from France, Germany,

Holland, Belgium and England. Many of these were concerned to show how effective their products were in minimising interference from the many transmitting stations then active in continental Europe. The number was large, almost 300 by 1934, and growing in power to reach a total of 4600 kW in that year, providing much more severe reception problems than experienced in Britain [15]. The designs exhibited were exclusively superheterodyne and often incorporating facilities for an antiparasitic aerial or an extra valve stage to accommodate this. This was a feature not often found in British designs. A simple version was to use a high doublet aerial with a twisted pair of unscreened down-leads. Interference induced in the two wires forming the down-leads cancelled each other out in the transformer with which it was terminated before the signal was delivered to the receiver input. Another arrangement was to include an extra 'parasitic' valve to which was connected an ordinary aerial and down-lead to the grid of the valve and a similar auxiliary aerial connected to the cathode, achieving a similar interference cancellation with the output of the valve passed on to the receiver input.

In addition to receivers a number of manufacturers were showing their valves, coils, loudspeakers and other component parts. The French valve manufacturers had by 1938 begun to follow American practice, and the range of valves shown all had metal envelopes. A few years earlier Hans Vogt had shown his new development of iron dust-cored inductances at the Berlin Radio Show of 1932, and at Marseille a number of examples of these new 'high Q' radio frequency coils were shown. Other new developments in France were press-button tuning using a motor drive, automatic frequency control, and a miniature display or 'magic-eye' tube used as an aid to tuning.

12.7 'Britain can make it' and 'Enterprise Scotland', 1946–1947

Before the extraordinary national exhibition of the Festival of Britain was to be seen in London on the anniversary of the 1851 Great Exhibition, two small but extremely popular exhibitions were to be promoted by the Council of Industrial Design in postwar Britain to '... *demonstrate the leadership of British goods in design'* [16]. Whilst the London Exhibition was concerned with industrial design on a fairly small scale, the Scottish Exhibition, which took place in Glasgow, emphasised the role of design in ship building and heavy industry.

'Britain can make it' was a demonstration of the strength and inventiveness of British Industry despite the single-mindedness of the recent industrial war effort over the preceding six years. One and a half million people visited this small Exhibition, confined to an exhibition space of only 8370 m^2, and every day queues formed at the entrance for admission, at the Victoria and Albert Museum in London. Accompanying the Exhibition was a conference sponsored by the Federation of British Industries. The papers read were concerned with the designer's place in industry and the design of machinery.

The Exhibition hall and the 24 allocated rooms in which the 5000 exhibits were presented were used to show the items grouped in specially designed settings, separate halls or in named sections, such as 'War to Peace', which showed design, materials and production methods arising out of wartime experience. Many of the exhibits related to design in the domestic field. Of significance were the 'wireless' receivers including an early portable model which could be slung over the shoulder 'like a camera', with the aerial incorporated in the carrying strap (using miniature valves but with fairly heavy batteries). This was described as '... *a noteworthy example of the application of war-time research to peace-time use*'. Other examples covered such diverse subjects as demonstrations of the packaging of goods for tropical climates, railway equipment design and machine tools. As with the later Festival of Britain, the Exhibition was intended to instruct the visiting public with such displays as the use in domestic products of materials, such as cast-iron, copper, aluminium and the newer war-time metals, magnesium and lightweight alloys.

12.8 Festival of Britain, 1951

Not since 1925 had Britain made any significant contribution to public exhibitions either at home or abroad. As the centenary of the Great Exhibition of 1851 drew near, consideration was given to mounting another large heterogeneous exhibition to be held in London in 1951. Governmental authority and, more importantly, finance were assured. In place of Prince Albert, the driving force for the 1851 exhibition, Herbert Morrison, Leader of the House of Commons in the postwar labour government, undertook a similar role in 1948 to plan the event. He was assisted with an extremely able committee, including Ian Cox as the Festival's Director of Science and Technology. A site on the south bank of the river Thames was chosen, ideally situated for road and rail communication and adjacent to Waterloo terminal railway station. The site was not large (11.1 ha) but extremely well planned to permit visitors to circulate freely around the exhibition (see Figure 12.5). Unlike the 1851 Exhibition, this was to be a national event, designed only to display 'British achievement in art, science, architecture and industry' [17].

Figure 12.5   The site of the 1951 Festival of Britain
(Central Press Photos)

A narrative approach was taken in the planning of the event. The site was divided into three main sections: 'Land of Britain', 'People of Britain', and 'British Discoveries and Invention'. Hungerford railway bridge formed a natural dividing line between the three sections, with the larger science, engineering and invention exhibits located to the west of the bridge.

Additional exhibitions on the theme of science and engineering also took place at the same time at the Science Museum in London, on the Festival Ship 'SS Campania', and at Kelvin Hall in Glasgow, which provided an exhibition of industrial power. The Glasgow Exhibition was of particular interest to electrical engineers since it provided a clear statement of the British electrical engineering industry at the midpoint of the 20th century. The latest equipment was shown in the hall of hydroelectricity, prepared by the Scottish Hydro-Electric Board. A 1/10 model of the Parsons 50 MW turboalternator was shown together with switchgear operating at a level of 2500–3500 MVA, supplied by Reyrolle, English Electric and Bruce Peebles. A demonstration was given of the BTH 350 A mercury-arc rectifier. Much industrial equipment was shown including industrial heating, radio-frequency induction heating, control electronics and portable electric tools. In the light current sector examples of up-to-date technology in short-wave therapy equipment, X-ray units, marine radar, direction finding equipment and measuring equipment were seen [18]. As part of the educational content of the Festival in Glasgow, attention was directed to work of the early engineers in a 'Hall of the Future', with effigies of Watt, Trevithick, Faraday, Parsons and Rutherford, and accompanying displays showing their contributions to engineering.

In the main site on the South Bank considerable interest was shown in a huge Telecinema auditorium, seating 400 people, arranged jointly by EMI and BTH. The display equipment was designed to show both conventional cinema films and transmitted television material using projection television and stereophonic sound. At the time, during the early days of British television, this was considered by some as the way forward for public entertainment, replacing the cinema, which had been so popular since the 1930s. The television unit employed a 50 kV tube with the picture projected through a Schmidt optical system, to ensure a brighter image, onto a 6.5 m wide screen. The definition was, however, limited to the 405 line television transmissions then in operation, and the projection tube proved to have a short life. Apart from its public demonstration at the Festival it never entered the commercial arena. At the time of the Festival the first provincial BBC television station had just been inaugurated at Sutton Coldfield near Birmingham, operating via a microwave link from London. A number of relay towers were necessary to ensure 'line of sight' transmission, and details of these towers and their equipment were shown by GEC, the designers of the link.

One of the features of the South Bank was the Shot Tower, originally used for the manufacture of lead shot, and in 1951 used to mount a disc-shaped aerial for a radio telescope. This 10 m diameter aerial was rotated by remote control from the outer space section of the 'dome of discovery', a large free-standing exhibition hall seen in Figure 12.5. By means of this control visitors to the Dome were able to see and hear radio waves transmitted from the sun and stars and also to witness a demonstration of the transmission of pulses of radio energy from the tower which were reflected from the moon's surface back to earth.

Held at the same period as the Festival of Britain, the Royal Society of Arts mounted an 'Exhibition of Exhibitions' in their rooms. This was opened by Princess Elizabeth, President of the Society. It showed several different types of exhibitions and the part played by the RSA in promoting them since their first exhibition in 1756, when the original purpose was stated as *'to show producers how they might improve their production with better machinery and implements'* —a surprisingly modern attitude. The story of the 1851 and 1862 exhibitions was given with many illustrations taken at the time, together with other exhibitions in which the RSA co-operated, both at home and abroad.

12.9 Ideal Home Exhibitions

The Ideal Home Exhibitions, sponsored by the Daily Mail newspaper, comprise the earliest series of exhibitions to consider domestic matters [19]. The first exhibition was held in 1908. At this time electrical domestic appliances, apart from lighting, were few and far between. In the 1910 Exhibition a number of heating and cooking devices began to be shown and included an electric knife and vacuum cleaners by the British Vacuum Co. Some of the advantages of electrical control were beginning to be realised in, for example, the control of gas appliances with an electric contrivance which turns on the gas and ignites it electrically [20].

With the establishment of the national electricity grid system in the 1930s, electrical appliances were featured more widely, permitting the demonstration of a number of kitchen and labour-saving appliances. Some of these were shown for the first time in Britain at the Ideal Home Exhibition: the first electric washing machine, electric storage heater, electric blanket and electric water softener [21]. Several were incorporated in the all-electric 'house of the future' at the 1928 exhibition and included an electric geyser, washing and ironing machines, a dishwasher, electrical heated blankets and an 'ultraviolet bath'. The living room was equipped with a radio receiver (and transmitter!), an electric typewriter, telenewsprinter and an 'automatic secretary'. This theme of 'the future' was to be repeated in many Ideal Home Exhibitions held since then, but none attempted to anticipate the common use of an aerocar, housed in the garage and 'designed to run out and rise vertically in the air', as suggested at the 1928 exhibition [22].

There was a significant measure of trade promotion in these exhibitions. In 1933 the Baird Television Company was arranging daily demonstrations of their receivers from a 30 line transmitter contained in the hall. In 1949 the Electrical Development Association mounted daily demonstrations of how the householder could convert an 'old-fashioned' kitchen into a modern 'all-electric' kitchen by stages, to incorporate some of the latest improvements in household appliances. This would include a refrigerator and an electric iron, which were beginning to be considered essential items for the modern home.The growing popularity of the domestic washing machine was demonstrated and, for the first time, a vacuum cleaner with an attachment to convert this into a floor polisher.

An interesting and unusual feature of the 1951 Exhibition at Olympia was a large replica of the 1851 Crystal Palace, measuring $32 \text{ m} \times 10 \text{ m} \times 15 \text{ m}$ high. This was presented in some detail and shown as an engineering accomplishment, being constructed entirely in magnesium alloy, first developed during the war for aeronautical purposes. The huge model only weighed between 3 and 4 t.

A feature of the yearly exhibitions has been the information stands provided by the public services—for example, the Post Office, telephones, etc. and the various ministries. At the 1965 Exhibition the Post Office was to demonstrate its mechanisms for sorting and transmission of letters in which electrical machines played an important role.

As seen at the other major public domestic exhibition, the National Radio Show, a large measure of entertainment was included within the exhibition hall. The BBC and ITV displayed television equipment with the newest receivers —although these were not to be compared with the range and detail to be seen at the Radio Show held in the same venue at another time of the year. In 1963 a feature which was to be repeated annually was a series of full-sized 'model houses' on display. Nine full-size houses were constructed within the exhibition hall. These were completely furnished with all the latest equipment and attracted a constant stream of visitors. They included the latest design of electrical kitchen appliances, with staff available to demonstrate them. The number of staff involved at these early exhibitions was very great. By 1965 the exhibition staff numbered over 10 000, employed mainly in demonstrating and answering queries. Much interest was shown by the public in those displays featuring work in the kitchen, which often included cooking demonstrations.

New ideas were continually being tried out in these 'model houses', such as arranging all the living accommodation on the first floor with the garage, control heating plant and other services in a central stem on the ground floor (which included a goods hoist). The upper floor consisted of four living rooms set in the form of a cross supported by cantilever arms radiating from the top of the central stem [23]. Warm air ducted heating feeding grilles in the living room was becoming popular and no less than five of the nine houses had this feature in the 1963 Exhibition. The electric versions used elements rated at up to 9 kW, which would be expensive by today's standards.

The practice of releasing new ideas in technology for demonstration to the public has been a feature of the Ideal Home Exhibitions. As recently as 1992 British Telecom demonstrated a prototype videophone for use in the home, which has yet to reach the commercial market [24].

12.10 The Electrical Association for Women

The growth in the demand for electricity for domestic purposes has already been remarked on earlier in this Chapter. A corresponding need for education in the use of electrical appliances for use in the home was recognised by the Women's Engineering Society and led to the formation of the Electrical Association for Women (EAW) in 1924 [25]. A major aim of the EAW was to explain the benefits of domestic electricity to ordinary women, and it did this through an educational programme, lectures, demonstrations and numerous exhibitions. The Association put much of the emphasis on travelling exhibitions. These appeared at agricultural shows in very many parts of the country throughout the period from 1930 to the mid-1970s in venues such as the Royal Show, the Great Yorkshire Show, the Royal Highland Show, the Three Counties Show and many others. In addition, city displays were held at home economics exhibitions in Bristol, Leeds, Leicester, Nottingham, Harrogate, Romford and at the Ideal Home Exhibition in London. Elements of these exhibitions included clear explanations of appliance plug connections and the safe operation of domestic equipment, such as cookers, fires, food mixers, washing machines, etc. [26].

In 1936 at a Bristol location, the EAW demonstrated an 'electric house' fitted with all the possible uses of electricity in the home and opened this to the public. This was before the days of the three-pin square plug, now standardised throughout Britain, and it is interesting to note that the plug and socket recommendation was the use of the Wylex all-in system, manufactured in Manchester, which became a *de facto* standard for many years in the north of England. On the occasion of its 21st birthday in 1945, the Association promoted a Women's Electrical Exhibition at Dorland Hall in which not only were domestic appliances shown and demonstrated, but the EAW paid tribute to the work carried out by the women's services in the WRENS, ATS, WAAF and in communications over the previous five years [27].

By the early 1980s the EAW had fulfilled its purpose: the safe and efficient use of electricity had become an accepted part of educated modern knowledge and it disbanded its organisation in 1986.

12.11 National Radio Exhibitions

The growth of public entertainment through the medium of radio, and later television, generated a huge desire amongst the public to learn more about these technical wonders and, of course, to see what industry could provide for installation in their own homes. Whilst to some extent this could be provided by agents of the individual radio companies, the need was felt and quickly satisfied for large public exhibitions whereby a number of companies' products could be compared side-by-side and the transmitting authorities could show what sort of facilities and broadcast material they were able to transmit. Radio exhibitions were started almost simultaneously in most major European countries and in the United States. Foremost amongst these were the comprehensive exhibitions arranged in Britain and Germany. The exhibitions in Britain were arranged initially by the Radio Manufacturers Association (RMA), which became the British Radio Equipment Manufacturers Association (BREMA) in 1945, and held yearly in London or Birmingham. They were directed almost entirely towards the general public, who attended in large numbers. From the first they included a generous measure of entertainment, with variety shows prepared by the BBC, taking place in the exhibition hall.

12.12 The British National Radio Exhibitions, 1926–1964

The first British National Radio Exhibition opened at Olympia, London in September 1926. Television was included for the first time in 1936. There was a gap in the yearly exhibitions between

1939 and 1947. The last 'radio show', as it was popularly named, in 1964.

The entire radio industry's year revolved around this exhibition, with the need to prepare the new receiving sets in time for public presentation. Its duration varied from 8 to 10 days and was attended by about 150 000, reaching its peak in 1934 with about 238 000 visitors. The number of exhibitors varied from 50 to 150, and not all of these were radio or component manufacturers—the Post Office and the BBC had substantial exhibits. The BBC also contributed a star-studded variety show, broadcast daily from the exhibition theatre, to which the public were invited.

Much of the early interest in radio was focused on home-constructed sets, made from kits of parts (it was often considered extravagant and slightly effete to purchase a set!). Cossor was one of the first companies to exhibit and sell its kit of parts for the 'Melody-Maker', a three-valve receiver, in the 1927 Radio Show (see Figure 12.6). Eventually up to $\frac{3}{4}$ million sets were sold to home constructors. Other manufacturers quickly followed this lead. The public looked to and expected technical advancements to be made at each succeeding exhibition, and were seldom disappointed. Up to the late 1920s most sets were operated by batteries and rechargeable cells. By 1928 several manufacturers were able to show an 'all mains' receiver, incorporating the new valves having indirectly heated cathodes which had been developed by Metropolitan–Vickers a few years earlier. The superheterodyne receiver, although introduced at some of the early exhibitions, did not make its mark until the 1932–1934 shows, when it became almost essential to achieve better selectivity to combat the proliferation of radio stations in Europe. This led to the decline in home construction, and subsequent exhibitions had fewer kit sets on show. Car radio was shown in 1934, following the lead established by America in 1930, when the first car radio was exhibited at the 1930 convention of the American Radio Manufacturers' Association.

It was, however, the coming of television that was to change the format and meaning of the yearly exhibition, although it retained its popular name of 'Radiolympia' for a number of years after the war. Baird was to show his rotating disc television receiver as early as the 1928 show, *'and was prepared to take orders for the set at the show'*, but little came of it [28]. Television for the public was not to be exhibited seriously until the 1936 show, just before a public service was initiated from Alexandra Palace in November of that year. Television did not entirely steal the show in those early prewar years, and we find, for example, a major innovation at the 1938 show to be the emphasis on

22   ADVERTISEMENTS   THE WIRELESS WORLD   OCTOBER 31ST, 1928.

BUY ALL THE PARTS

£7·15·0

for the wonderful NEW COSSOR 'Melody Maker'

FOR £7. 15. 0. you can buy all the parts for the wonderful new Cossor Melody Maker. They are sold in a sealed box—sealed to prevent substitution of inferior or unsuitable components—sealed to ensure your obtaining the right parts and those only. Everything is included—the handsome all-metal cabinet, the three Cossor Valves, the wire and even the simple tools. There is nothing more to buy. Get to know all about this wonderful new Cossor Melody Maker.

◄ Fill in this coupon NOW !

Please send me free of charge one of your Constructive Envelopes which tells me how to build the new Cossor "Melody Maker."

Name

Address

Advt. A. C. Cossor. Ltd., Melody Department, Highbury Grove, London. N.5

Mention of " The Wireless World," when writing to advertisers, will ensure prompt attention

Figure 12.6   The Cossor 'Melody-Maker', at the 1927 Radio Show
(Bussey, 'Wireless; the crucial decade', 1990)

push-button tuning, made possible by the development of automatic frequency control.

By 1936 domestic television consisted only of cathode-ray tube receiver designs, generally using electrostatic deflection, which necessitated deep and bulky cabinets. To improve the receiver presentation many manufacturers made use of vertical mounting for the tube, an angled mirror forming the top of the cabinet against which the picture could be seen. The screen size was large when compared with that obtained in the rotating-disc receivers, and 30.5 and 38 cm diameter tubes were not unusual. At the following show, in 1937, Philips were showing their innovative projection tubes and large-screen television, 51 × 41 cm, but few receivers were sold due to the short tube life and consequent expensive replacement.

By 1939 cathode-ray tubes using magnetic deflection became available, permitting shorter tubes to be produced and a more compact cabinet design. At this time television sales were up to 19 000 per year. The young and flourishing industry then came abruptly to an end with the outbreak of war, when transmission ceased for almost seven years.

National Radio Exhibitions recommenced with the 15th exhibition in 1947 to coincide with the resumption of television transmissions. At its first postwar show it was seen that big strides forward had been made, particularly with Pye Ltd., who showed an advanced receiver, the B16T, using 'line-flyback EHT generation', enabling the bulky mains transformer to be reduced in size and a lightweight receiver to be manufactured. This was not fully achieved, however, until the following year when it became possible to arrange the valve heaters in series so that with a later model, the B18T, shown in Figure 12.7, Pye ushered in the 'transformerless' television receiver, which became a standard arrangement for all subsequent manufacturers.

At this first postwar exhibition the hall at Olympia was shared not only with television and the new radio receivers but also with an industrial sector which enabled the public to see some of the equipment developed during the war years, such as Metropolitan–Vickers radar equipment, Marconi direction finders and blind landing apparatus, an STC 100 kW short-wave transmitter and, possibly the most significant device for the future, a small BTH germanium rectifier, heralding the beginning of the semiconductor revolution.

Radio receivers provided their own innovations. Flywheel tuning appeared and advances were made in high-fidelity audio equipment. Although the combined radio and gramophone receiver made its debut at the 1933 radio show, its popularity was never very great until the advent of microgroove records in 1950. The radiogram, and the portable gramophone capable of playing these new records, were shown with great success.

Figure 12.7   The Pye B18T 'transformerless' television receiver, 1948
(Courtesy of the National Museum of Photography, Bradford)

The 1949 exhibition was the first to display projection television receivers to the public as a successful commercial venture. Several manufacturers exhibited their new sets, all incorporating the Mullard 6.35 cm diameter tube working with an applied potential of 25 kV. The tube was mounted vertically and an angled mirror directed the picture on to a vertical screen. This was generally 38 × 31 cm, considerably larger than that possible with direct viewing.

The 1950 radio show opened in a new venue, the new International Exhibition Centre near Birmingham, built in 1949 and first used for the annual meeting of the British Association for the Advancement of Science. This venue was chosen for the radio show in order to give people in the Midlands an opportunity to appreciate the second BBC television service which had just commenced operation from Sutton Coldfield. Ninety firms exhibited, compared with 140 at Olympia the previous year. There were no outstanding changes in television, apart from the short-lived appearance of dark-tinted screens to improve daylight viewing and the replacement of the 23 cm screen size by one of 30 cm. At the previous (1949) Exhibition the Pye company had provided, somewhat prematurely, a demonstration of colour television over a closed circuit using the RCA system of three separate colour tubes and a rotating colour filter. The scheme was subsequently abandoned in the United States and nothing was seen of this at later exhibitions.

The 1951 Exhibition was to emphasise the advantages of 'miniaturisation' in radio equipment, and a number of ranges of sub-miniature valves with wire-ended connections were shown, suitable for use in hearing aids and electronic equipment.

By the late 1950s the transistor had just emerged from the laboratory to replace the thermionic valve in many applications, and its use in portable radio receivers at the 1956–1957 radio shows proved immensely popular—indeed the lightweight portable receiver became known at the show, and for some time afterwards, as 'the transistor', an incorrect but popular appellation. The process of assimilation of transistor development into radio design continued in 1958 when printed circuit wiring was seen for the first time. This enabled yet smaller sets to be fabricated, particularly in car radio receivers, which now operated directly from the car's 12 V battery. For the first time stereophonic amplifiers and gramophone equipment reached prominence, establishing a growing market for the new stereo micro-groove records. The average size of the television screen at this show was larger, with many of the new 43 and 53 cm diagonal rectangular tubes being shown. This was made possible by the advent of 'wide-angle' tubes, having scanning angles of up to 90 deg. These larger tubes led to the demise of the projection-tube receiver which was no longer to be seen at this and subsequent exhibitions.

In 1964 the new high-definition standards for television transmission on an ultrahigh frequency (UHF) enabled a 625 line transmission to take place on a second channel. The earlier 405 line transmissions continued to take place, however, and new 'dual standard' switchable sets were shown at this show. The '405/625 switch' in these receivers was of great complexity since it was required to change not only the line repetition frequency but also the receiver band tuning and consequently affected most of the set functions (and the price!). These complex receivers did not remain long on the market, however, but their successors were not exhibited at the National Radio Exhibition. The last public exhibition organised by BREMA was in 1964. Some individual efforts were made to sustain the public's interest. An outstanding example was the Pye/Echo travelling train, hired from British Rail. This consisted of 13 coaches, two of which contained a local television station, and which toured a number of cities in Britain during September and October in 1965. The industry had become fragmented, however; several companies no longer supported the Association and many disagreements on colour standards became evident between the Association and the BBC prior to the establishment of a colour service in 1969. The imposition of resale price maintenance on the trade and import of foreign radio and television sets all contributed to the lack of cohesion within the industry, and the corporate will to mount future exhibitions vanished [29].

12.13 The German Radio Exhibitions

Similar radio exhibitions were also held in the United States and Germany. The German procedure was to divide the exhibition into two parts, one for trade and one for the general public, acknowledging a dichotomy in presentation which we consider in the following Chapter. German radio exhibitions had a special section devoted to television and, unlike the British exhibitions, a diverse number of television systems were shown working together under the auspices of the German Post Office. In 1931 these included mirror wheel, Nipkow disc and cathode-ray tube systems. The number of lines comprising the picture could be 48, 60, 84, 90 or 100, with aspect ratios varying from 3:4 to 5:6 and with different grades of definition. The first radio exhibition in Berlin during 1929 exhibited television, using a Nipkow scanning disc, just a year after Baird had shown similar equipment at the 1928 National Radio Exhibition in London. However, in the case of the Berlin event this was a two-way exhibit given by the German Post Office to demonstrate the possibilities of television–telephone communication [30]. Two public booths were set up at opposite ends of the stand, to allow visitors to see and hear each other simultaneously when making a call from one booth to the other. A single 30-hole scanning disc rotated at 12.5 rev/s was used at each end, but the picture was poor and the continual traversal of a spot of light across the face of the speaker extremely distracting.

By 1936 the Berlin Radio Exhibition was giving a demonstration of 375 line television with interlaced scanning, the use of electronic cameras for real-time outdoor transmission and an early version of projection television [31]. Substantial firms such as Fernseh A G, Telefunken, Loewe, Lorenz, Philips and Tekade were showing receivers operating on 180 line definition (with a few advanced receivers at 375 lines, 25 pictures per second) from ultrashort-wave transmissions. Most of the receivers used cathode-ray tubes, some having the large diameter of 45 cm. Interest in high definition receivers was heightened at that time due to the 1936 Olympic Games, held in Germany, from which some outdoor events had been televised using a Farnsworth type of electron camera, manufactured by Fernseh A G. Telefunken made use of their Iconoscope camera to demonstrate closed-circuit television. Mechanical receivers using a Nipkow disc or a mirror drum were still being manufactured, and several were shown. Of particular interest was the sophisticated projection receiver by Tekade which used a 180 line receiver and a rotating mirror drum with an arc lamp for illumination. A picture of about 60 cm square, said to be 'of reasonable brightness' was shown. A telephone-television service was inaugurated between Leipzig and Berlin in the same year [32].

The British Radio Exhibitions ceased to be held just before the BBC commenced their stereo broadcasts on VHF in 1966. West Germany with its network of stations had already started to broadcast in stereo and in 1964 the German Radio Exhibition (held in Stuttgart that year) was to show stereo receivers from almost all their set manufacturers. Television at this exhibition showed a parallel development of the use of transistors to that being carried out in Britain but had not yet reached the stage of an 'all-transistor' colour television receiver. The German firm of Graetz & Telefunken did, however, exhibit a 65 cm viewing screen, the largest yet shown in a public exhibition. Further development of cassette type recorders were shown by Philips and Deutsche Grammophon, including one for car use (a narrow tape cassette receiver by Philips was first exhibited at the 1963 Berlin Radio Exhibition). Several other tape systems were shown but were all plagued at that time by the lack of a common agreed standard for recording. Three television tape recording schemes were shown (the term 'videotape' had yet to be invented) by Loewe Opta, Philips and Grundig, all reel-to-reel systems and fairly massive in construction [33]. With the cessation of the British Radio Exhibition in 1964 the West German Radio Exhibition became the premier location to view the comparative development of domestic radio and television receivers on a regular basis, although a number of smaller private exhibitions continued to be held from time to time in most countries of Western Europe.

12.14 References

1 'Handbook of Electricity Supply Statistics' (Electricity Council, London, 1990)
2 'Hampstead Electrical Exhibition', *The Electr. Rev.*, 7 Oct. 1910, **67**, p. 579
3 'Guildford Electrical Exhibition', *The Electr. Times*, 8 Oct. 1928, **74**, p. 553
4 'A corkscrew railway', *Sci. Am..*, 1924, **130** (2), pp. 88–89

5 'Wembley British Empire Exhibition; Guide to the Palace of Engineering' (HMSO, London, 1926)
6 'British Empire Exhibition—broadcasting the King's speech', *The Electr. Times*, 1 May 1924, **65**, pp. 498–499
7 'Jigger': 'British Empire Exhibition', *The Electr. Times*, 29 May 1924, **65**, pp. 633–634
8 'Electricity in service', *The Electr. Times*, 29 May 1924, **65**, p. 635
9 'Wembley British Empire Exhibition: Handbook of the Exhibition of Pure Science' (HMSO, London, 1926)
10 'South Africa's Empire Exhibition', *The Engineer*, 9, 16 Oct., 13 Nov. 1926, **70**, pp. 372–373, 406, 511–513
11 'The Empire Exhibition, Glasgow', *The Engineer*, 26 Aug. 1938, **166**, pp. 225–262
12 'Light and lighting at the Empire Exhibition, Glasgow', *Light & Lighting*, 1938, **31** (4), p. 95
13 GUERRIER, L.: 'L'electricité dans les Expositions régionales', *Electricien*, 1939, **55** (1707, 1708), pp. 194–200, 218–225
14 'The Ideal Home Exhibition', *The Electr. Rev.*, 11 April 1930, **106**
15 BUSSEY, G.: 'Wireless—the crucial decade 1924–1934' (Peter Peregrinus, London, 1990), Chap. 3
16 'Britain can make it', *Engineering*, 1946, **87**, pp. 303, 443
17 'Engineering at the Festival', *Engineering*, 13 April 1951, **171**, p. 421
18 'The exhibition of industrial power, Glasgow 1951', *Engineering*, 3 Aug. 1951, **172**, p. 133
19 'Sunshine and leisure', *Engineering*, 12 Mar. 1965, **199**, p. 352
20 'The Ideal Home Exhibition', *The Electr. Times*, 14 April 1910, **37**, p. 365
21 'Domestic water softener', *Engineering*, 20 April 1928, **125**, p. 483
22 'Ideal Home Exhibition', *The Electr. Times*, 1928, **73**, pp. 319, 387–388
23 'Ideal village', *The Engineer*, 29 Mar. 1963, **197**, p. 464
24 'BT demonstrates videophone', *Br. Telecommun. Eng.*, April 1992, **11**, p. 70
25 SYMONS, L.: 'The Electrical Association for Women 1924–1986', *IEE Proc. A.*, May 1993, **140** (3), p. 140
26 Institution of Electrical Engineers Archives, NAEST, Savoy Place, London
27 IEE Archives, NAEST93, 1993
28 BURNS, R.W.: 'British Television—the formative years' (Peter Peregrinus, London, 1986)
29 GEDDES, K., AND BUSSEY, G.: 'The Setmakers' (BREMA, London, 1991)
30 Editorial: 'Television at the Berlin Radio Exhibition', *Telev. J.*, Oct. 1929, **1**, p. 382
31 TRAUB, E.H.: 'Television at the 1936 Berlin Radio Exhibition', *J. Tv. Soc.*, Dec. 1936, **1**, Pt. 3, pp. 182–187
32 BURNS, R.W.: 'Prophesy into practice: the early rise of videotelephony', *IEE Eng. Sci. Educ. J.*, 1995, **4** (6), pp. 33–39
33 'German Radio Exhibition', *Wireless World*, 1965, **71**, pp. 495–498

Chapter 13

The modern era

13.1 World fairs and trade fairs—a new dichotomy

Towards the end of the interwar years it had become apparent that the whole international exhibition movement had begun to change in a fundamental way, to show a dichotomy which reminds us of the difference between the English and French ideas of the exhibition early in the 19th century (see Chap. 2). On the one hand world fairs had diverged to become largely festivals of entertainment and education, often on a grand scale, whilst the trade fairs had become purely technical and industrial, attracting buyers, sellers, designers and others to a specialist exhibition devoted to one aspect of technology—a process that was accentuated after the Second World War

   In both cases the ease and rapidity of air travel had made it essential that the language difficulties of visitors arriving at these fairs needed to be addressed. Since the 1970s simultaneous translation of aural presentations began to appear to support the multilingual text material that had been available for some time. At trade fairs it had become necessary to have one or two representatives of companies exhibiting, to be fluent in two or more European languages, although English has, to some extent, assumed the status of a common language among engineers at these events. For general visitors to world fairs the street and direction signs now take a standard form, shown in Figure 13.1, and adopted at all succeeding fairs arranged under the auspices of the Bureau International des Expositions (BIE see below). Finally, computer software, available on the interrogation screens, now an essential feature of all large fairs, began to be available.

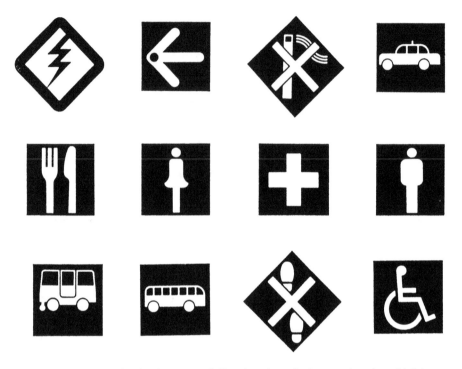

Figure 13.1　Standardised street and direction signs for international world fairs

In this and the following Chapter we consider how the two forms of the exhibition have developed, commencing with some of the larger world fairs arranged recently in Europe, America and the Far East.

Before doing so, however, the organisational aspects of international world fairs will need to be discussed. Mention was made in Chap. 10 of the efforts of the French in establishing a unifying organisation, L'UCAD, initially to rationalise the French contribution to exhibitions at home and abroad since 1851. Some 80 years later this was to influence the formation of an international organisation, 'Bureau International des Expositions' (BIE), with its present offices in the Avenue Victor Hugo in Paris. Initial moves towards internationally agreed control of large exhibitions were made by Henry Cole and other foreign commissioners to the Paris Exposition of 1867. They proposed that the scope and duration should be controlled and exhibitions rotated between the various capitals, with exhibits arranged by class rather than nationality. A formal meeting took place at an International Convention in Belgium in 1912, signed by 15 countries, with a further meeting in 1928, when 43 nations participated which resulted in a 'Convention Regarding International

Expositions'. This led to the formation of the BIE in 1931, with the United States joining in 1968, bringing the number of participating countries up to 48. From 1988 the BIE has formally recognised two forms of exhibitions: a registered exhibition having a duration of between six weeks and six months, and a recognised exhibition having a duration of between three weeks and three months. The interval between two registered exhibitions proposed from any country must be at least five years and only one recognised exhibition may take place between two registered exhibitions. Both types of exhibition are to have a specified theme and in the case of the smaller recognised exhibitions their total surface area must not exceed 25 ha. A set of rules has been set out governing the rights and obligations of exhibitors and the position regarding customs regulations and other matters [1]. The rules regarding intervals between exhibitions have not, however, been adhered to in all cases, and it remains to be seen whether this will be modified in the future.

13.2 Brussels World Fair, 1958

The Brussels World Fair was the first major exhibition on an international scale to be held after the cessation of hostilities. The Festival of Britain was, of course, a predominately national event. After holding successful national exhibitions in Liège and Antwerp (1905 and 1930), Guent (1913) and Brussels (1910 and 1935) the Belgian Government planned the Brussels Universal and International Fair in 1958 on a theme of 'International Humanistic Development', but it could equally have been termed the 'Nuclear Energy Exhibition' since it was the first international exhibition to describe fully the postwar activities in the harnessing of nuclear power for electricity generation.

The Fair was situated some 6 km from Brussels on a 200 ha site (now a public park within the present-day extended city). The peaceful use of atomic energy in the 1950s was of great public interest with Europe leading the way in the development of nuclear power stations. To illustrate this a major exhibit at the Fair was conceived as a gigantic model representation of a crystal molecule of iron, and known as the Atomium (Figure 13.2). It consists of a series of huge interlinked spheres, 18 m in diameter, each one big enough to house scientific exhibits or a restaurant. The spheres were linked together with escalator tubes or walkways, symbolising the binding nuclear forces. Within the spheres and tubes connecting them, exhibits and display boards described the principles of nuclear energy and its use for

Figure 13.2   The Atomium at Brussels World Fair, 1958
(Official Guide, Brussels World Exhibition, 1958)

electrical power generation. Some 45 nations took part in the Fair, with major contributions from Britain and the United States [2].

Many national contributors stressed the promotion of nuclear power to meet their future energy needs. Most impressive was the British contribution housed in two large pavilions with the collective theme of 'power for progress'. In one of these was shown a huge model, on 1/12 full scale, of a complete nuclear generating station incorporating a gas-cooled graphite moderated reactor. The model was chosen to depict the reactor building at Hunterston, one of the first of four commercially built nuclear power stations then under construction [3]. Details of the British contribution show the advanced state of this nuclear reactor programme compared with those in other countries [4]. Also shown in the British pavilion was a display of the current research then being carried out at Harwell on fusion, which had culminated in ZETA—the Zero Energy Thermonuclear Assembly—at that time showing great promise as a future generator of electricity [5]. A one-third scale model of ZETA was shown with a simulating interior flash at 10 s intervals to represent the pinched discharge within a torus in which a temperature of $5 \times 10^6 \, °C$ had previously been attained in 1957.

A consortium of Belgian industries displayed models of reactors scheduled to begin operation in 1960. The French Government described their plans to use nuclear power to meet at least 25% of France's energy needs by 1967, a plan which was more than accomplished as the century progressed. Other nuclear exhibits were shown in the United States pavilion with a model of a pressurised water reactor and, in the Soviet pavilion, details of their proposed breeder reactor programme and a model of a nuclear power station. In addition to the plans for fixed nuclear reactors a model of their nuclear-powered ice-breaker, Lenin, was displayed. The American Argonne Laboratory exhibited tools for the new atomic age: an electronically controlled manipulator, a 'mechanical hand' for handling radioactive materials shown to the public for the first time, using a stereo television display for observation, and a 100 channel pulse height analyser.

Technology was very much in evidence throughout the exposition, much of it developed since the 1940s. Apart from the Festival of Britain this was the first public exhibition in which the tremendous advances in technology generated during and since the war years were being put on display. The second British pavilion stressed the advances made in electronics and other technologies during and since World War II, with demonstrations of radar and thermionic valve production. This included a working exhibit of a marine radar scanner, simulating actual conditions afloat. Standard Telephones and Cables displayed one of their submarine two-way telephone repeaters of the kind that had been recently installed in the new Anglo–Belgian submarine cable system, having a sea length of ~102 km. It provided 120 two-way telephone channels between London and Brussels. With the industrial displays was exhibited the newly developed Crusilite silicon carbide electric furnace, capable of operating at a temperature range extending to 1500 deg C. Part of the exhibition grounds was devoted to the use of electricity in transport by rail, sea and air. Electric traction was shown with full scale and model exhibits of electric and diesel-electric locomotives. Of particular interest at the time was a scale model of a P & O liner currently being built in Belfast, which was being equipped with electric propulsion machinery. Power and light supplies for the British pavilions were controlled by two substations, having a capacity of 1000 kVA, installed by English Electric and located within a glass-walled compartment serving as a working exhibit.

In the Russian pavilion the first spacecraft, Sputnik 1, was shown— perhaps the most exciting technological exhibit at the Fair. Also exhibited was a magnetic tape system controlling a three-dimensional milling machine—the first time such a device had been shown to a public outside specialised trade fairs.

In one of the largest industrial pavilions IBM introduced visitors to the new world of electronic calculation, with early data processing equipment and an electric typewriter. Bell Telephones were also displaying a computer, *'capable of thousands of calculations in a few (sic) seconds'*. Words new to the public at that time: automation, cybernetics, semiconductor, transistor, etc., were given meaning with appropriate demonstrations. A working transistor amplifier and oscillator were shown by Bell and a ferrite core memory by Philips of Eindhoven. Philips also had its own pavilion at the Fair. This was designed by Le Corbusier and used 2000 slabs of prestressed concrete, each of a different shape, suspended between steel cables to form a freely curving hyperbolic structure. Some 300 loudspeakers were built into this curved surface with a battery of amplifiers presenting a 'poem electronique' with sound, light and an ever changing projection of various colours playing on the inside surface [6]. This was the first of a flood of such sensation-seeking synthesis of sound and light in coming international exhibitions. The Swiss pavilion was showing an atomic clock (a novelty at this time), which made use of an ammonia molecular oscillator at 24 GHz as its standard in a frequency control circuit involving a Reflex-Klystron tube, with a quartz crystal to provide a secondary reference. Additionally at a number of stands and pavilions within the Fair were shown the new radio and television receivers, radar installations, an automatic telephone exchange and an example of an electronically controlled lathe. A number of major exhibits concerning electrical and hydraulic power were displayed in the main hall, shared by a number of participating nations.

The Brussels World Fair was presented at a time when there was a dearth of information available to the general public on the immense strides taken by technology during the war years and in the decade that followed. To some extent the gap was filled by the Festival of Britain, but this exhibition was somewhat parochial and did not represent events on a world stage as were seen at Brussels. It was a remarkable exhibition and judged a great success by all the countries who participated, as indeed may be seen from the attendance of over 41 million visitors from all parts of Europe and America.

13.3 Seattle 'Century '21' Exposition, 1962

This was the first major exhibition to be held in the United States since 1939. It was also the first major exhibition affected by the controlling activities of the newly constituted BIE, whose approval was sought and obtained in spite of the much larger New York Exhibition already proposed for 1964 (within the two year ban on similar exhibitions being held in the same country). An earlier Exhibition in Seattle in 1909 sought to draw attention to the situation of Seattle as a gateway to the Pacific and Alaska and to its industrial potential. A similar theme was initially proposed by urban planners for the 1962 Exposition. However, in the aftermath of the Russian launch of the Sputnik satellite, which demonstrated the strength of Soviet science and engineering, the American Government in Washington considered it opportune to use the occasion of the Seattle Exposition to demonstrate the industrial strength of America as a whole and the far-reaching ideas developed by the United States since the end of the 1939–1945 war. Consequently the Exposition became a predominantly American event and generous federal assistance was forthcoming to the Seattle Exposition organisers. Incidentally, both the 1909 and 1962 Expositions left lasting legacies for the city of Seattle. The 1909 event cleared land that was later to become the site for Seattle's University, and the 1962 Exposition was planned to provide several permanent buildings to be used as a cultural centre and a permanent science exhibition. The site for the 1962 Exposition was not large in comparison with other international exhibitions (30 ha). The Exposition planners had been impressed by the Festival of Britain in 1951, arranged on an even smaller site, and tried to improve the quality and interest for the visiting public in much the same way.

The theme chosen was opportune for the time, 'Man in Space', and its major purpose to depict life in the 21st century. The dominant architectural structure at the exhibition was a 182 m tower, the Space Needle, topped by a revolving restaurant (Figure 13.3), a monorail commuter service from downtown Seattle, a United States Science Pavilion and the 'Coliseum Century '21', depicting a possible lifestyle in the next century.

The largest of the Exhibition areas was given over to the house of science. It consisted of five sections in separate buildings:

Building 1: *House of Science,* which included a display consisting of six simultaneous 35 mm projectors with commentary showing, in a $1\frac{1}{4}$ h performance, the activities of science in all its variety

Building 2: *History of science,* which aimed to introduce four divisions of science research, the electromagnetic spectrum, the structure of matter, genetics and the concept of the universe

Building 3: *Spacearium,* a huge planetarium inside a 24 m diameter geodesic globe

Building 4: *Methods of science,* the largest exhibit, separated into 27 sections, each showing how scientists work in a given discipline

Building 5: *Horizons of science,* emphasising that advances in science cannot be expected to provide final answers.

The whole of this presentation was very much technology-controlled using the aids provided by electronics and control equipment in a way not possible at earlier exhibitions.

The science area was attractively lit at night (see Figure 13.4), and accounted for a sizable fraction of the loading for lighting and exhibitions at the Seattle Exposition. The total requirement was some 22 600 kVA, derived from the city's own power stations. It was expected that the major buildings on the Seattle site would eventually form part of a permanent exhibition site for the city, 'The Pacific Science Center', and for this reason permanent underground cabling was installed from the 120 000 kVA substation several blocks away from the site.

A monorail service ran from central Seattle to the Exposition, powered by a 600 V DC supply from the city's transport system (it runs still, as a feature of the modern city of Seattle) [7]. Monorail schemes proved popular with the exhibition-going public in the 1960s. The

Figure 13.3   The Space Needle at the Seattle World Fair, 1962
(*National Geographic Magazine*, 1962, 122)

Figure 13.4   Lighting at Seattle World Fair, 1962
(US Government Publication)

Turin Exhibition, held the previous year, was one of the first to make use of this form of mass transport, with a 1.5 km track traversing the Exhibition area. Later exhibitions, in New York (1964), Montreal (1967), Osaka (1970) and the Louisiana World Exhibition in 1984, all installed monorail systems. They were eventually overtaken by electro-magnetic levitation systems, as a visitor attraction, ushered in by the International Transportation Exhibition in Hamburg in 1981, which shuttled passengers at 90 km/h along a 1 km track supported on high pillars. This was followed by the Tsukuba Expo'85 which demonstrated a high-speed magnetic-levitation train travelling at 300 km/h.

13.4 New York World's Fair, 1964–1965

The Fair was planned to commemorate the 300 year anniversary of the founding of New York City with the theme of 'peace through understanding' and constructed in Flushing Meadows, on the same site as the 1939–1940 World Fair. The Fair was privately sponsored. Its symbol, the Unisphere, a large model of the Earth, replaced the trylon and perisphere of the 1939–1940 Exposition. Except for Spain, Austria, Denmark and Greece, the major European countries did not officially take part, although private European exhibits from Belgium, Berlin, France, Sweden and Switzerland found a place in the grounds. It was, however, significant for the large number of eastern countries represented. Almost all the East Asian countries took part, from India to Japan, with Japan making a strong presence.

To enable visitors to traverse the 260 ha site a large monorail encircled the grounds. This consisted of two-car trains, each carrying 80 passengers, automatically controlled on two parallel closed loops 12 m above the ground. Each car was driven by two 10 hp motors coupled to the wheels via an eddy current clutch and brake [8].

A number of electrical firms had their own pavilions, which in some cases were very extensive. General Electric made a feature of electricity in the home. It arranged a large stage within its pavilion such that four auditoriums revolved around the stage area, showing the history of electrical developments in the home over several decades.

General Motors repeated their view of the future, in an improved form as 'Futurama II', which was viewed by passengers seated in a continuous chair-chain, as in the 1939–1940 New York World Fair. In this case, however, 463 cabins were employed, each seating three passengers, with two passenger conveyors, one to enable visitors to board the continuously moving chair-chain at the beginning of the ride, and the other so that they could leave safely at the other end [9]. Passenger conveyors or 'moving pavements' were first used at the 1893 Chicago World Fair and again at the 1900 Paris Exposition. Both of these were much longer than those seen at the 1964–1965 New York World's Fair, where they were used as a means of access to the pavilions rather than as a means of transport. Some 26 passenger conveyors were installed at the 1964–1965 Fair for use by various chair-chains throughout the grounds.

The Bell Telephone pavilion made use of four of these, and several others were used to serve rotating theatres. Each theatre was a closely co-ordinated system in which the sizes of the seating areas had been determined by the speed at which the conveyor ramps could load them. For example, the electric power and light pavilion contained a turntable with seven chambers, each holding 175 people. Passenger conveyors were used to funnel batches of 150 people swiftly into each of these viewing areas in turn. The design of all these passenger conveyors was similar to that described for the 1900 Paris Exposition in Chap. 10 with minor modifications, such as rubber covered driving wheels, and generally containing only one moving platform. The Bell Company also demonstrated here for the first time a 'touch-tone' telephone dialling system.

The Fair included much entertainment material, with use made of Walt Disney audio-animation techniques throughout the site. The IBM pavilion, a huge ovoid building mounted on stilts, provided a multiple screen cinema in which 17 projectors made use of film, slides and live presentations to show how computers process information. The General Electric pavilion, one of the largest in the Fair, showed for the first time in a public demonstration the functioning of a controlled nuclear fusion, currently the subject of much research and many hopes in the American nuclear industry. In keeping with the idea of mass public entertainment, as well as providing a measure of education, the Bell Company presented a moving chair ride giving a tour of communications history.

Despite these attractions the Fair was not a success with the general public. Cheap and tawdry commercialism was given its head, with predictably disastrous results. The site suffered from the effects of audible and visible pollution. Throughout the grounds, background music from the central distribution system clashed with individual pavilion sound-tracks, and the thousands of makeshift signs vied with the few official ones to create a cacophony of sound and vision [10]. This was not to occur in subsequent international expositions recognised by the BIE.

13.5 Expo'67, Montreal, 1967

The Montreal Universal and International Exposition of 1967 was conceived as a celebration for the centenary of the establishment of the Canadian Confederation and was approved by the BIE as Expo'67 (all major international exhibitions have carried the appellation, Expo', followed by the year, since the late 1960s). The site at ile Saint Hélène was an interesting one, consisting eventually, after much civil engineering (half of the site was reclaimed from the St Lawrence River), of almost 405 ha, which included a number of lakes, canals and an artificial island of 116 ha [11].

Considerable attention was paid to transport {12]. A mass transit system started at the main gate, Place d'Accuil on Cite du Havre, and terminated on ile Saint Hélène. It consisted of eight computer-controlled trains of six cars each. It incorporated a full-sized heavy rail electric rapid transit system using cars of the type recently installed on the new Toronto subway system [13]. A feature of this system is that the cars ran on rubber tyres, which made it considerably quieter than competing systems elsewhere. Three smaller electric minirail systems passed through and around many of the Exposition pavilions. These were actually light monorail systems of novel design, first shown at a Transportation Fair held in Munich in 1965. In addition to these, two further electric mini-rail systems were transferred from the Swiss National Exposition of 1964, held at Lausanne, for use at the Montreal Exposition. These provided transport within the 55 ha amusement park, known as La Raide, situated within ile Saint Hélène. What was not transferred from Switzerland, and is an interesting adjunct to the Swiss minirail system, was the method of loading the passengers on to the railway carriages, which in Lausanne had run continuously around a complete loop of track. The loading mechanism took the form of a rotating station platform or turntable, almost 30 m in diameter, situated alongside the track. The passengers were required to enter the turntable via an overhead walkway and down a ramp to its centre and then to traverse to the edge where the speed is high enough to permit safe entry on to the moving carriages of the train, arranged to just skirt the turntable at that point and travelling at almost 5 km/h (reduced from its normal operating speed of 10.5 km/h between stations) [14].

The Montreal Exposition was well supported by 60 foreign national governments and three American states. The American pavilion was designed as a huge Buckmaster Fuller's geodesic dome shown in Figure 13.5. This was linked with other national pavilions by means of a monorail, seen in the illustration. Use of the dome was considered initially as a method of climatic control but was not found to be entirely successful. Expo'67 was notable for its imaginative use of several new film techniques, including a giant split screen with multiple images, later to become a common feature of many international exhibitions.

The theme for the event was 'man and his world', and was expressed by 17 subthemes having titles such as 'man the explorer', 'man the producer', 'man in control', 'resources for man', etc. With this type of classification the distribution of electrical products, development and engineering became rather diffuse. This was becoming true of all international exhibitions proposed and realised in the modern era. No longer were the public to be impressed by the bringing together of the means of generation, distribution and demonstration of the diverse uses of electrical energy in a hall of machinery or palace of electricity, as had been the case previously. Instead the generation was

Figure 13.5   Buckmaster Fuller's geodesic dome at Montreal Expo'67
(Allwood, 'The Great Expositions', 1978)

taken for granted and the applications of electricity found in almost every display.

13.6 Expo'70, Osaka, Japan, 1970

This was the first significant international exhibition to be staged by Japan and the first to be held in Asia. Its central theme was 'progress and harmony for mankind'. It was held to celebrate the 100th anniversary of Japan's entry into the Meiji period, in which Japan made the transition from a feudal, Chinese-oriented culture into a modern nation state. The Exhibition attracted 64 million visitors and ran for 180 days.

The site covered 250 ha in Suita City, a satellite city on the outskirts of Osaka in a region that was originally bamboo groves and rice fields. Since the site was so extensive the transportation arrangements were important and note was taken of the success of Montreal Expo'67, with their adaptive light railway systems. The scheme adopted in Osaka took the form of a 4.4 km computerised monorail system, similar to the successful scheme used in the 1964–1965 New York World Fair. In addition, battery-powered cars, each able to accommodate six people, were used together with a tubular moving pavement travelling at 2.5 km/h. To fill any gaps remaining in the transport requirements a fleet of electric bicycles, using nickel–cadmium cells, was available to visitors, whilst overhead 22 globular cable cars traversed the site.

Seventy-seven countries took part, together with four international organisations, all housed within a total of 85 pavilions—each country having one or more pavilions for their exhibits. The most striking building was the Festival Plaza. It consisted of an enormous transparent roof, resting on six gigantic pillar supports and covering an area of 32 000 m^2. The entire structure weighed 6000 t and rose 30 m at its tallest point. It was a superb engineering feat, somewhat akin to the 'dome of discovery' at the 1951 Festival of Britain Exhibition, but very much larger. An artificial sun within the dome shone down on to a lake and fountains within the building. The building was serviced with an air-conditioning system to maintain an 'atmosphere of spring' within the plaza, walled in by 'air curtains' to assist the control of the internal climate. This was not the only architectural innovation constructed at Expo'70. The British pavilion consisted of four exhibition halls suspended from steel shafts, each over 36 m high, whilst the United States pavilion was built partially underground, with an air-supported fibre glass dome covering an area of ~11 243 m^2. In the Japanese pavilion wave power was demonstrated with a device to convert the energy of artificially created 1 m high waves into a 500 W electricity supply.

As discussed earlier, electrical exhibits had ceased to be collected into any one area either by subject or nation and were to be found carrying out many different roles and in different sections of an exhibition. This was the case, for example, with computers, which are now widely applied to all facets of modern life. Thus in the United States pavilion the most interesting examples of electrical science were to be found in the space section, which contained the Mercury, Gemini and Apollo space capsules with their controlling technology.

13.7 Knoxville Energy Fair, 1982

The 1982 Knoxville International Energy Exposition was conceived on a theme, 'energy turns the world'. This was the first world exposition to present this theme since the Brussels Exposition of 1958. Knoxville was certainly a suitable location for demonstrating this kind of activity. It was located within 32 km of Oak Ridge's National Atomic Laboratory and its American Museum of Science and Energy and in the centre of the Tennessee Valley Authority with its vast hydroelectric generators. With over 11 million visitors it proved to be a popular 'theme fair' in the United States. The site was small (29.2 ha) and situated next door to the campus of the University of Tennessee. It featured two large 'signature' structures, a large wedge-shaped building for the United States pavilion and the 'sunsphere', which comprised a 5-storey global structure mounted on a tower with a total height of 81 m.

Most of the corporate exhibitors displayed their exhibits in a 9000 m² 'technology and lifestyle centre'. America's electric energy exhibit comprised exhibits from General Electric, Tennessee Valley Authority and Breeder Reactor Inc., which combined to demonstrate the future role of electricity to include current developments in cooling and heating technology for homes and industry and a model of the fast breeder reactor at the nearby Oak Ridge Laboratory. Gas Exhibits Inc. used laser light channelled through transparent tubes to demonstrate how natural gas flows from an oil well to the surface. The Communications Satellite Inc. (COMSAT) displayed its 'satellite to home' technology. Dairymen Inc., representing Louisville's dairy farmers, showed for the first time their method of milk processing at ultrahigh temperature (UHT) to provide a long-life product, a technique that was soon to become universal.

Twenty-four nations took part, but many of the exhibits were far removed from the theme of the fair. The Egyptian contribution, for example, provided a display of models of artifacts taken from their ancient civilisation. The China pavilion made much of its replica of the Great Wall, terra cotta warriors and a restaurant with a chef from Beijing. It did, however, demonstrate a solar-powered dragon boat on Fort Loudoun Lake and an installation for the production of methane gas from farm waste. Several exhibitors were concerned to display alternative sources of energy. Australia contributed a series of windmills pumping water to irrigate their display of rain forest environment and demonstrated the value of solar power generation for living in the Australian outback. Canada exhibited several alternative sources of energy contrasted with Ontario's powerful nuclear energy program, which benefited from Canada's plentiful uranium deposits. Japan displayed their latest computer-controlled walking robots that talked about energy in several languages. Saudi Arabia presented huge solar energy collectors and West Germany models of its nuclear reactors. The United States Pavilion contained a number of levels, each generously supplied with touch control video screens. At one level a whole wall was covered with 20 of these concerned with aspects of the nation's current energy problems (this was at a time of controversy for the future of nuclear energy in the United States). The pavilion also provided a computer-driven interactive display activated by the visitor, who, by touching the screen, could select the speaker or subject desired and experience a 2 min 'sound and vision bite' on the subject. An IMAX wide-angle film was also shown giving a description of America's energy resources and technology. The educational content was undeniable but proved a little indigestible. One commentator considered that it would take an average visitor about 6 h to complete a tour of the exhibits in this pavilion alone! [15].

The Fair also included a considerable entertainment content: a circus, rides, river boats, singing groups, shops, fireworks, football, baseball, a 'family fun area', and a Baptist pavilion devoted to 'spiritual energy', in line with the theme of the Exposition. As may be seen from an advertisement of the time (Figure 13.6), the entertainment value of a visit to the Fair was significant and had much in common with the New York Fair of 1964–1965, both far removed from the ethos of earlier American fairs, such as Chicago 1933 and New York 1939–1940.

13.8 Expo'85, Tsukuba, Japan, 1985

Japan staged its second major international exposition at the new Science City of Tsukuba, situated some 50 km north-east of Tokyo. The intention of the Japanese Government is to localise much of their modern high technology in specially created cities, such as Tsukuba and 18 further 'science cities' planned throughout Japan in the coming decades to stimulate technological innovation. It comes as no surprise, therefore, to find technology high on the list of activities at this Exposition. This carries the theme of 'dwellings and surroundings—science and technology for man at home' and '... *is an attempt to acknowledge both the technological and the humanistic requirements for a good life in the decades to come*' [16]. Despite the emphasis on technological innovation the organisers favoured a policy of not announcing new inventions as such. This was in deference to a widely held view that inventions had their 'down side' as well as advantages (e.g. environmental ones) where technology was present but only in its supporting role. This was underlined by the aim expressed at its opening of '*creating a scientific and technical vision of the 21st century that will contribute to young people's understanding of the world of science and encourage them to undertake careers in technology*'.

The Exposition covered a 100 ha site and consisted of 48 pavilions,

Get in on the adventure, thrills and discovery of The 1982 World's Fair.

If you miss it this year, you've missed it forever!

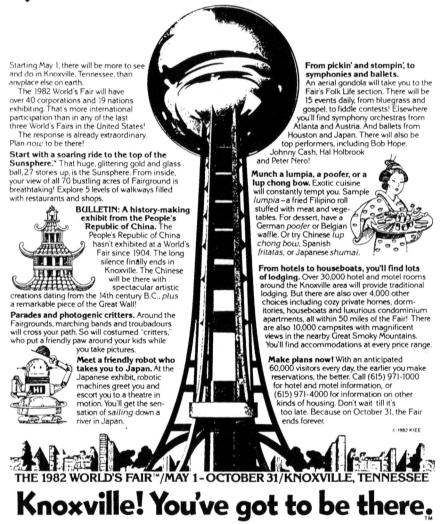

Starting May 1, there will be more to see and do in Knoxville, Tennessee, than anyplace else on earth.

The 1982 World's Fair will have over 40 corporations and 19 nations exhibiting. That's more international participation than in any of the last three World's Fairs in the United States!

The response is already extraordinary. Plan *now* to be there!

Start with a soaring ride to the top of the Sunsphere. That huge, glittering gold and glass ball, 27 stories up, is the Sunsphere. From inside, your view of all 70 bustling acres of Fairground is breathtaking! Explore 5 levels of walkways filled with restaurants and shops.

BULLETIN: A history-making exhibit from the People's Republic of China. The People's Republic of China hasn't exhibited at a World's Fair since 1904. The long silence finally ends in Knoxville. The Chinese will be there with spectacular artistic creations dating from the 14th century B.C., *plus* a remarkable piece of the Great Wall!

Parades and photogenic critters. Around the Fairgrounds, marching bands and troubadours will cross your path. So will costumed "critters," who put a friendly paw around your kids while you take pictures.

Meet a friendly robot who takes you to Japan. At the Japanese exhibit, robotic machines greet you and escort you to a theatre in motion. You'll get the sensation of *sailing* down a river in Japan.

From pickin' and stompin', to symphonies and ballets. An aerial gondola will take you to the Fair's Folk Life section. There will be 15 events daily, from bluegrass and gospel, to fiddle contests! Elsewhere you'll find symphony orchestras from Atlanta and Austria. And ballets from Houston and Japan. There will also be top performers, including Bob Hope, Johnny Cash, Hal Holbrook and Peter Nero!

Munch a lumpia, a poofer, or a lup chong bow. Exotic cuisine will constantly tempt you. Sample *lumpia* – a fried Filipino roll stuffed with meat and vegetables. For dessert, have a German *poofer* or Belgian waffle. Or try Chinese *lup chong bow*, Spanish *fritatas*, or Japanese *shumai*.

From hotels to houseboats, you'll find lots of lodging. Over 30,000 hotel and motel rooms around the Knoxville area will provide traditional lodging. But there are also over 4,000 other choices including cozy private homes, dormitories, houseboats and luxurious condominium apartments, all within 50 miles of the Fair! There are also 10,000 campsites with magnificent views in the nearby Great Smoky Mountains. You'll find accommodations at every price range.

Make plans now! With an anticipated 60,000 visitors every day, the earlier you make reservations, the better. Call (615) 971-1000 for hotel and motel information, or (615) 971-4000 for information on other kinds of housing. Don't wait till it's too late. Because on October 31, the Fair ends forever.

© 1982 KIEE

THE 1982 WORLD'S FAIR™/MAY 1 – OCTOBER 31/KNOXVILLE, TENNESSEE

Knoxville! You've got to be there.™

Figure 13.6   Advertisement for the Knoxville International Energy Exposition, 1982
*(Reproduced with permission from World's Fair Mag., 1982, **2**)*

some allocated to specific countries and others to commercial companies, all of which had a free hand in their design. Some of these were:

Japan's theme pavilion:	a large measure of education directed to young people on the 'humanistic role of science', showing scientific advances geared to the situation of people in the developing countries
US pavilion:	a major attraction was a description of how the United States was designing a fifth-generation computer
Sony Jumbotron:	a giant colour television screen, measuring 25 m high and 40 m wide (it needed to be viewed from a distance of at least 200 m)
Fuyo robot theatre:	featured a robot that could read music and play the piano, and another that could produce a line drawing of a visitor in minutes and then asked, 'Is it like you?'
History pavilion:	covering general world history rather than the development of technology and science

and:

UK pavilion	TKD pavilion
Shueisha pavilion	IBM pavilion
Sumitomo 3-D fantasium	Steel pavilion
Mitsubishi pavilion	Electric pavilion
KDD Telecomland	Gas pavilion.

Japan Air Lines demonstrated a high-speed surface transport (HSST)—a mass transit system which makes use of electromagnetic levitation to suspend the cars about 1 cm above the rail. It used a linear motor for propulsion and was designed for a speed of 300 km/h. Unfortunately only 350 m of track were provided to link two of the pavilions, so that it did not contribute a great deal to the transport problems of the Exposition.

All the buildings of the Exhibition site are now incorporated as part of the growth of the 2700 ha area of Tsukuba Science City [17].

13.9 Expo'86, Vancouver, 1986

This was only the third international exhibition to be held in Canada, previous exhibitions being in Toronto in 1884 and Montreal in 1967. It was intended to celebrate the city's centenary but developed quickly into a lavish form of tourist promotion [18]. The site was small, 71 ha spread along Vancouver's shore line, against over 400 ha for the Montreal exhibition. Its theme, 'world in motion—world in touch: human aspirations and achievement in transportation and communication', apparently left room for the integration of extensive fairground booths, amusement rides and much entertainment amongst the pavilions of the Exhibition. Technology was represented with the space shuttle, a feature of the United States pavilion, and an emphasis on recent mass-transit systems, specifically in the Japanese electromagnetic levitation railway, seen earlier at the Tsukuba Exposition in 1985. Much that was being exhibited had been seen before at American and Japanese exhibitions a few years earlier. The presentation techniques, typically large curved screens—some three-dimensional and others as kinetic split-screens—although competently carried out, also emphasised the spirit of *déjà vu* considered by many critics to be evoked throughout the Exhibition. An unusual form of audio-visual presentation by General Motors showing early forms of communication, however, did make use of a new technique, that of holography, which had not appeared at earlier exhibitions. The most popular events at the exhibition were surprisingly high quality entertainments by La Scala Milan, Britain's National Theatre, Beijing People's Art Theatre, Japanese Opera and the Royal Thai Ballet.

Vancouver 1986 was one of the first exhibitions to become dominated almost entirely by an entertainment aim. Anderson comments somewhat scathingly that

> '... *theme park or carnival would aptly describe Expo'86. And despite the unusual 'first' of the combined presence of the United States, China, Cuba, and the Soviet Union, Expo'86 boasted nothing more substantial than hosting a lively party, complete with nightly fireworks'* [18].

In this and a number of later exhibitions the entertainment features owe much to the ideas promulgated by Walt Disney and his 'experimental prototype community of tomorrow' (EPCOT), which opened in 1982 in Florida, described as 'a permanent World Fair of sorts built and maintained by the Disney organisation' [20].

13.10 Expo'88, Brisbane, 1988

We noted earlier the change in presentation of international expositions that was becoming apparent since the early 1980s. The organisers were placing less emphasis on education and technical interest, either from motives of cost, or more probably in an attempt to attract a much wider range of visitors than at earlier international events. Since 1851 all large international exhibitions provided entertainment in some form or other and at the beginning of the 20th century this often took the form of a separate entertainment area or 'fun park', adjacent to or as part of the exhibition site. This function had begun to invade the main exhibition site and in some cases usurp the informative or technical content almost completely. With the Brisbane Exposition this process is seen to dominate to such an extent that publicity matter concerning the event was promulgated with

> 'A scale of entertainment not previously seen or experienced in Australia setting new heights for diversity and excellence during World Expo'88. Twice daily parades with scope and creativity to rival Disneyland wound through the site, while the River Stage, set on the Brisbane River, with the city lights as the night-time backdrop provided a unique venue for mass entertainment' [21].

The Exposition at Brisbane was arranged to celebrate the bicentennial year of Australia's foundation under the theme of 'leisure in the age of technology'. Technology was indeed used effectively to present an extravaganza in terms of architectural design and lighting throughout the comparatively small site. Robots were employed to welcome the visitors in 32 languages, stepping stones across cascades triggered decorative water jets, and high-speed rides through simulated space landscapes were presented. It was the first international exhibition designed especially for a television audience, not only in Australia but worldwide. As noted by the Entertainment Director in charge of the television arrangements,

> 'The production was transmitted to a potential Australian audience of 15 million, whilst television stations from most of the other 35 participating countries created a potential global audience of hundreds of millions'.

The opening day spectacle required a large entertainment cast together with a crew of some 9000 people to control transmission, recording, lighting, transport, crowd control, etc., and was described as

'...the merry-spirited opening setting the tone for a fun-packed six months of interest and entertainment'. The event continued with *'... more than 25,000 entertainment events presented twelve hours each day, seven days a week for the duration of the World Expo'.*

Previous expositions at Tsukuba and Vancouver had made wide use of electronic information systems for visitors, and in Brisbane these facilities were even further developed. Some 56 public information terminals (PIT) were distributed over the site. Each PIT (work station) contained a large database of text, audio and graphical images (both still and motion sequences), and all were made available to the visitor via a large colour touch-screen panel, seen in Figure 13.7. This user-interactive system enabled not only text to be available on the screen (in English and four versions of Japanese script, Kanji, Kastakana, Hiragana and Romaji) but also videodisc frames and moving images. All the information was updated daily and the PIT checked for maintenance requirements and daily statistics collection (e.g. frequency of access) [22].

The Brisbane Exposition of 1988 touched the nadir of World Fair presentation. Since the start of the 1990s, whilst the information

Figure 13.7   Information terminal at Brisbane, 1988
(*Aust. Telecommun. J.*, 1988, **38**)

technology features have remained, the trend towards domination by entertainment has been much reduced, and is considered further in Chap. 15, after we have looked at the parallel development of trade fairs in Chap. 14.

13.11 References

1 'The International Bureau of Expositions and Regulations respecting it'. The International Bureau of Expositions, 56 Avenue Victor Hugo 75783, Paris cedex 16, 1988
2 'Official Guide to Brussels World Exposition 1958' (Desclee & Co, Tournai, Belgium, 1958)
3 'Brussels 1958', *The Engineer*, 1959, **205**, p. 602
4 'Report from Brussels', *Engineering*, 1958, **185**, pp. 520–527
5 'The British Pavilion at Brussels'. Associated Newspapers Brochure, 1959, pp. 26–27
6 'Electronic Poem, Le Corbusier', *Int. Light. J.*, 1958 (3–4)
7 HERALD, G.: 'Underground Distribution Serves Fair at Seattle', *Electr. World*, 18 June 1962, **157**, pp. 36–37
8 'Monorail at the New York Fair', *Engineering*, 5 Feb. 1965, **199**, p. 179
9 SMITH, R.F., and KHORAM, L.: 'A design summary of the GM Futurama II ride at the 1964–65 New York World's Fair', *Gen. Motors Eng. J.*, 1964 (2nd qu.), pp. 2–10
10 'The Great Expositions' (Macmillan, London, 1978), pp. 163–164
11 'Expo'67: creating the site', *Engineering*, 1967, **204**, pp. 17–19
12 'Expo'67: transport systems', *Engineering*, 1967, **204**, pp. 423–426
13 'Comfort on the subway', *Engineering*, 1970, **209**, p. 251
14 'Internal transport systems at the Swiss National Exhibition', *Railw. Gaz.*, 1964, **120** (9), pp. 363–364
15 KAHL, W.L.: 'The U.S. Pavilion at the Knoxville World Fair 1982', *World's Fair*, 1982, **2** (2), pp. K8–K9
16 TRACY, D.: 'Tsukuba Expo'85', *World's Fair*, 1985, **5** (1), pp. 5–9
17 ASHIHARE, Y.: 'Expo'85 architecture'. (Arch. Inst. Jpn., 1985), pp. 4–38
18 ANDERSON, R., and WACHTEL, E.: 'The Expo Story—Vancouver 1986'. Seattle, 1986, p. 20
19 FINDLING, J.E., and PELLE, K.D.: 'Historical Dictionary of World's Fairs and Expositions 1851–1988' (Greenwoood Press, NY, 1990)
20 MEIKLE, J.L.: *World's Fair*, 1982, **2** (2), pp. 2–6
21 WALSH, A., and HALL, J.: 'World Expo'88, Brisbane' (World Expo'88 Media Centre, Brisbane, 1988)
22 CLARKE, J.: 'Expo'88 Brisbane', *Telecom J. Austr.*, 1988, **38** (1), pp. 63–68

Chapter 14

Trade fairs

14.1 The modern movement—trade fairs

The idea of a purely commercial trade fair is not new. Indeed the original purpose of a Fair was simply a place to exchange or sell goods, as we discussed in Chap. 1. What is new is the scale of trade fairs seen since the middle of the 20th century, far removed from the limited local fairs set up, for example, by the French in the 1800s or, in content, from the combined educational and entertainment intention of the present world fairs or expositions. Not that the trade fair is without an educational content, but this too has become specialised and directed towards a series of particular commercial products rather than to the broader field of instruction which had become typical of the technology-related exhibitions of the late 19th or early 20th century.

Smaller modern trade fairs had been promoted in many countries for some considerable time. The Leipzig Fair, first held in 1165, flourished in the middle ages and by the 17th century began to be held in substantial premises, frequented by well-to-do merchants (Figure 14.1). It became an international fair and a centre of book-selling and printing in the 19th century to continue as a major fair up to the outbreak of World War II. The Fair recovered remarkably thereafter and held its first postwar exhibition in October 1945 to become in the 1960s the principal showplace for engineering design from Eastern Europe. Some 40 countries were represented, with a major contribution made by East Germany, Poland and Russia. Unlike many modern trade fairs the Leipzig Fair covers a broad spectrum of the engineering field and represents development in light current electronic engineering as well as heavy machinery and plant.

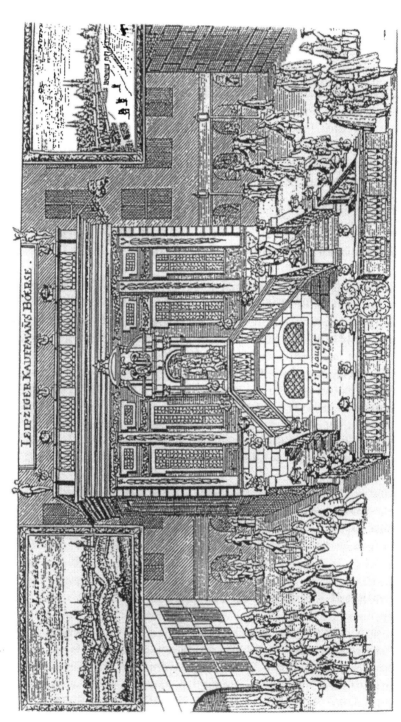

Figure 14.1   The Leipzig Trade Fair in the 17th Century
(Reproduced by permission of E.A. Seemann, Leipzig)

Television, for example, has been covered since 1936, when it provided the first public demonstration of television outside Berlin by the opening of the Berlin–Leipzig two-way television link. The wideband link was significant in that it made use of coaxial cable for the first time [1].

Other smaller fairs in Lyons and Bordeaux, in their modern form, have been held since the beginning of the 20th century and, in a revival of the Paris Fair, since 1917. Before the British Industries Fair was established in 1915 (see below) several electrical exhibitions were mounted for the trade in London and Manchester. The London events took place at Olympia in 1905, 1908 and 1911. Sir William Preece, then President of the IEE, was closely associated with these and opened the 1911 Exhibition. All aspects of electrical engineering were represented, from generators to domestic electrical appliances, and this exhibition was significant for the showing of the first really successful metal filament lamp in Britain by GEC known as the 'Osram' lamp. GEC also demonstrated an electrical 'Punkah' fan said to *'give an exact electrical reproduction of the 'flick' of the celebrated "Punkah-wallah"'* [2].

Recent trade fairs abroad include two which made their appearance in the second half of the 20th century: the Milan Fair, which in 1960 featured extensively the nuclear industry in the wake of the 1958 Brussels Exposition; and St Erik's Fair in Stockholm, a yearly fair which is concerned mainly with industrial equipment and electrical engineering. The Brno International Trade Fair in Czechoslovakia rivalled Leipzig in its display of engineering exhibits, particularly machine tools. It has been held since 1967 and attracts visitors from all countries in Europe. In the United States international trade fairs did not commence until 1950, when a fair was held in Chicago specifically to assist western European countries in selling their products in the American market with the revival of European producers in the aftermath of World War II.

14.2 British Industries Fair

The British Industries Fair (BIF) was perhaps most typical of the smaller modern trade fair development and had been held yearly in London and elsewhere since 1915. The first BIF, supported directly by the Government, was held in the Royal Agricultural Hall, London. It originated from a wartime need to replace the goods of foreign manufacture, both industrial and domestic, whose import had suddenly ceased on account of the War. British manufacturers responded enthusiastically to remedy these specific shortages, and 591 exhibitors at this first exhibition showed samples of their replacement wares to over 30 000 visitors to the hall [3]. Its original motive was lost after a few years being replaced by the opportunity to '... *enable home and foreign buyers to examine all classes of British manufacture*', and as a consequence it ceased to command general public interest, being essentially a 'buyers' exhibition' for modern engineering and technology requirements.

The BIF remained a national exhibition and was held each year at a variety of venues in London and provincial towns. It made use of the White City in Shepherds Bush, the site of the 1908 Franco–British Exhibition, for a number of years before moving to the Crystal Palace at Sydenham in London, before this was unfortunately burnt down in 1936. From the 1930s the BIF was held in Castle Bromwich, Birmingham, as well as London, with the Birmingham location used for engineering exhibits. Several new developments were shown in Birmingham, including the first serious attempt in 1938 to display the technical features of electric road vehicles for the export market. Messrs A.E. Morrison exhibited a range of small 500, 600 and 1000 kg electric delivery vans. A 4.5 hp motor was fitted giving a maximum road speed of 30 km/h supplied from a 30 cell battery. Speed was controlled by a foot pedal operating a series of switches to cut in or out series resistance, which led to rather low efficiency. The GEC displayed a range of domestic appliances such as cookers, electric fires, hot plates, fan heaters and irons, rivalling those seen at the Marseille Exhibition held in the same year (Chap. 12).

1938 and 1939 were remarkable years for trade exhibitions; 40 were held in Europe, with 16 in the spring buying session of 1939. During the war years only the Swiss Industries Fair remained active.

No BIF exhibitions were held between 1939 and 1947. Restarting the BIF after the war was stongly supported by the Department of Trade, who recommended that a permanent building should be erected to house this and similar exhibitions [4]. In the event the postwar fairs were to take place at the exhibition centres in Olympia and Earls Court in London and the National Exhibition Centre in Birmingham. The exhibitions were usually opened for 11 days, with admission by trade ticket and with the public only admitted after 1600 h and on Saturdays.

In the first BIF after the war in 1947 the British electrical industry showed examples of radio-frequency heating, welding equipment, electric trucks and heavy electrical plant, such as Ferranti transformers with ratings of 5000 kVA. Some domestic electrical equipment, electronic control equipment and radar were shown. A number of new inventions received their first public showing at the BIF, an example being a new electronic weighing scale by Avery in 1953. Computing equipment began to make its appearance, with a display of British Tabulating Company's office and accounting equipment, Powers Samas punched card machines and Hollerith punched cards. Digital computers were featured at an early stage in their development, and the 1956 Fair was responsible for the first public appearance of the new Ferranti 'Pegasus' computer which was claimed to be *'very simple to program'*. The Ferranti 'Mercury' was also shown at this fair, the first computer to use 'floating point' arithmetic and representing a big step forward in computer design. These were both valve computers; computers using transistors were becoming available in 1956 and the first of these, the Metrovic 950, was shown not at the BIF but at a smaller exhibition, the Instruments, Electronics and Automation Exhibition held in London in 1957 (Figure 14.2) [5].

The BIF was to last 42 years, providing a much needed showcase for British engineering. In 1958 the President of the Board of Trade and

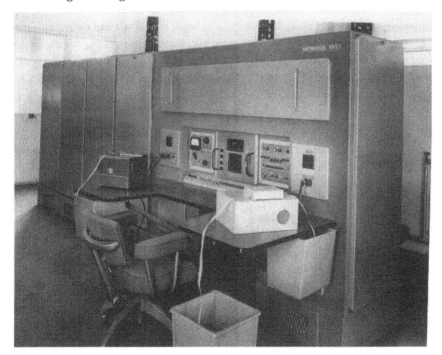

Figure 14.2 The Metrovic 950 Computer at the Instruments, Electronics and Automation Exhibition in 1957
(Courtesy of the National Archive for the History of Computing)

Industry in the United Kingdom stated that, following a meeting with the Federation of British Industries, the vast majority of trade associations no longer wanted to contribute to a general engineering trade fair such as the BIF. Instead they favoured the type of 'single industry' fair which was becoming popular in the UK and abroad [6]. The last BIF was held in 1958 [7]. This was confined to heavy engineering and hardware at Castle Bromwich in Birmingham, and since government funding had by then been withdrawn, was financed by the Birmingham Chamber of Commerce.

14.3 The Electrical Engineers' Exhibition (ELECTREX)

An important exhibition for the electrical industry is the annual Electrical Engineers' Exhibition, promoted by the Association of Supervisory Electrical Engineers (ASEE), and held since 1951[1.] The first exhibitions were small but successful in creating a climate of interest for exhibiting the products of a single industry, and in 1953 the first ASEE exhibition was held in Earls Court Exhibition Centre (Figure 14.3), which was to become its home until it moved to the National Exhibition Centre in Birmingham in 1976 [8]. In these early days it competed to some extent with the BIF and, like the BIF, started to exhibit digital technology in 1956. New developments in germanium and silicon power rectifiers were shown by BTH, and digital computers by Ferranti. These latter included their Mercury and Pegasus computers, which carried new features of floating-point arithmetic, magnetic core stores and magnetic drum storage.

By 1957 it was the largest exhibition devoted to a single technology to be held at Earls Court in London. A number of co-operating organisations took part in this Exhibition for the first time: the Illuminating Engineering Society, the Central Electricity Authority, the Post Office, the BBC, the Royal Navy and the Royal Air Force.

A major theme for this Exhibition was training and education in industry. This had been part of the aims of the Association since its inception, marked by the presentation of awards for the best industrial product, the best domestic labour saving equipment and the award of bursaries in lighting engineering. At its opening ceremony, Lord Hailsham, then Minister for Education, used the occasion to

---

[1]'The Association of Supervising Electricians', as it was called at its foundation in 1914, became 'The Association of Supervising Electrical Engineers' in 1927. The organisation of exhibitions has been arranged by a separate company, 'ASEE Exhibitions Ltd', since 1952

Figure 14.3   *The Electrical Engineering Exhibition at Earls Court, 1966*
(Courtesy of ASEE Exhibitions Ltd)

announce a £70 million technical education programme. An educational display was presented by the Central Electricity Authority, which exhibited detailed models of Bradwell and Berkeley nuclear power stations, which had just been commissioned.

The Electrical Research Association (ERA) showed its progress in electrical surge generators producing up to 130 kW of energy for the testing of insulators and contact breakers. Some of the results of this work were shown by Ferguson Pailin of Manchester with its new 132 kV air-blast circuit breaker. At this fair a new feature shown was a single mercury lamp, rated at 4.75 kW, powerful enough to light up the whole 31 000 m^2 of the exhibition hall from a central location. It was claimed that 36 such lamps would be sufficient to light an area of 1.82 ha [9]. In the domestic area Siemens Edison Swan Ltd showed their new moulded telephone handset, the first of its kind to incorporate printed circuits, and one of the early examples of personal paging equipment for staff location in a medium sized organisation.

The contribution of the co-operating organisations at the ASEE was sometimes considerable, and in 1959 the golden jubilee of the Illuminating Engineering Society was noted with a form of entertainment combining light, colour, stereophonic sound and animation along the lines of a *son et lumière* in Greenwich Park.

After 1962 the ASEE, in conjunction with the organisation representing most of the electrical exhibitors, BEAMA, decided to hold their exhibitions biannually and to make their exhibition an international one, inviting contributions from electrical manufacturers abroad. By 1965 the exhibition was attracting exhibits from 70 overseas exhibitors. New techniques continued to be shown. In 1968 the first production of aluminium foil transformers was shown— perhaps the most radical change in basic construction of the distribution transformer since silicon steels for the transformer core were introduced in the 1890s. In 1970 the exhibition became known as ELECTREX and recognised the growing importance of electronics to the electrical power engineer. Automation and control of electrical drives were featured, with an emphasis on thyristor-controlled drives, then a fairly sophisticated method of control.

The value of ELECTREX as an international exhibition was augmented in 1980 when agreement was reached with its corresponding electrical exhibitions in the mainland of Europe, ELEC in Paris and INTEL in Milan, to co-ordinate its exhibition dates, a procedure which led to the setting up of a joint body INELEX to maintain this co-ordination. Finally in 1992 the scope of exhibits at ELECTREX was extended to include power generation, transmission and distribution, with the name of the exhibition changed once again to ElectroTech to emphasise this new role.

The 1996 ElectroTech exhibition at the National Exhibition Centre in Birmingham covered five halls and included power generation & distribution, control & instrumentation, energy utilisation, environmental technology, lighting, generators and motors and many other electronic products. The lighting exhibits were enhanced by reappearance of the Lighting Industry Federation and the participation of many European Union countries.

14.4 Other British trade fairs

Specialised trade fairs, held on a yearly basis, have been a feature of industrial life in Britain since the end of World War II. They were given a fillip in 1946 through the publication of the Governmental Ramsden Report which emphasised the important role that exhibitions and trade fairs could play in promoting British goods abroad. The report commended the British industrial fairs of the past and urged that they should be revived. It was this report that encouraged governmental assistance in supporting British participation in a number of exhibitions abroad, such as the Brussels World Fair of 1958, Seattle Century'21 in 1962, Montreal Expo'67 and particularly the Festival of Britain in 1951.

One important trade fair that is held yearly, with considerable public enthusiasm, is the Society of British Aircraft Constructors' annual flying exhibition at Farnborough in England. This is an international exhibition with a portion of the ground dedicated to an exhibition of communication and radar equipment, details of which are reported in the technical press each year [10].

A yearly exhibition, established in 1944 for the radio and television industry, was the Exhibition of British Radio Components and Associated Equipment organised by the Radio and Electronic Component Manufacturers Association (RECMF) and held initially in an elegant location at Grosvenor House, Park Lane. This had an objective to

> '... *ensure a high standard of quality, design and workmanship and to promote standardisation of radio components and accessories*'.

It also served to assist designers of domestic radio equipment to compare alternative suppliers' components at a single location and helped to maintain a low cost for the complete receivers in a way which would be difficult to achieve easily from a manufacturing site.

The RECMF provided a venue to show to a wide audience some of the new ideas and inventions in electronic components as they became available commercially (often these had been shown earlier on an experimental or laboratory basis at the Physical Society Exhibition (Chap. 10)). For example, in the 1946 Exhibition, the first since World War II, the new high permeability alloys such as radiometal, mumetal, rhometal, permalloy, silco and ticonal made their first public appearance in the form of transformer stampings and other shapes which were then available for peacetime assembly work. The properties of the new selenium rectifiers were also shown at this Exhibition. In an interesting exhibit in 1947, Tungsram Valves were showing their new 'all purpose' valve, the UA55, incorporated in a prototype receiver, the first ever to make use of deposited metal wiring (an early version of the printed circuit) and with the receiver constructed entirely by a single production machine, Sargrove's

Electronic Circuit Making Equipment (ECME), which caused enormous interest in 1946–1947 [11]. In the 1950 Exhibition, Mullard Ltd showed for the first time production components made of their new ferrite material, 'Ferroxcube': radio inductors, television deflection coil assemblies and line output transformers. Initially the RECMF was conceived as a showplace for the British component industry and occasionally sponsored meetings in an overseas location, such as Stockholm in 1961, to further British industry. In 1965 it widened its appeal by becoming an international exhibition at a new venue in Olympia. After 1966 it merged with three other technology exhibitions to form the International Instruments, Electronics and Automation (IEA) Exhibition. The IEA had been founded in 1950, initially sponsored by the Scientific Instrument Manufacturers Association, which held its first exhibition at Olympia. This, as its title suggests, was concerned with the 'light current' electronics field rather than the heavy electrical engineering of the BIF or the ASEE. Components for automation were exhibited, significantly now including controllers, analogue computers and, towards the late 1950s, digital computers. These latter included computers for accounting systems as well as for general-purpose scientific calculation, showing an overlap with other, more specialised computer exhibitions, which began to appear at about this period. The 1957 IEA exhibition demonstrated an extensive range of available commercial computers, including machines from Metropolitan–Vickers (model 950), Standard Telephones and Cables (Stantec–Zebra) and the National Cash Register Company (model 405). Several hundred firms took part—almost 1000 by its sixth exhibition in 1966, when it attracted 102 000 visitors, of whom about a third were from abroad. In 1976, the IEA, which had by then moved to the National Exhibition Centre (NEC) in Birmingham, combined with the ASEE to promote a trade show covering a wider range in electrical engineering than either of the two exhibitions taken alone. This arrangement lasted until 1982 when the IEA ceased to participate in annual trade fairs.

14.5 Modern international world trade fairs

The modern world trade fair is fully international and extensive, taking place over very many exhibition halls—functional buildings having few or none of the architectural features found in world fair sites. As we saw from the reasons expressed for the demise of the BIF, current business interests since about 1960 have generally considered to be best served by limiting the contents of a fair to one particular range of technology—sometimes a very specialised trade. Entrance is often restricted to trade buyers and those from participating industries. Unlike world fairs and other nonspecialist exhibitions, they are usually of short duration (two to six days) and held at the same location and time of the year. This is ideal for business personnel attending such events, enabling them to reach a large number of buyers of a particular range of items in a short time. A feature of these trade fairs, which has become increasingly important, is their use as a springboard for innovative or new products. Often, a new idea, which later develops into a major product range distributed on a worldwide basis, is seen first at these exhibitions. Although since 1950 trade fairs have been a feature of industrial life in North America (the first was held in Chicago in 1950 for the special purpose of assisting the rehabilitation of western European countries in their postwar overseas trade), nearly all the large trade fairs today are held on a regular basis in Europe, although incorporating multinational exhibitors from all parts of the world.

Several of these having a strong electrical basis are discussed below, commencing with the Telecommunications Exhibition held under the auspices of the International Telecommunications Union in Geneva every four years.

14.6 International Telecommunications Union

This is located in Geneva, Switzerland, as the United Nations agency for worldwide telecommunications, and supported by over 164 countries[2]. Its origins go back to 1895, when it was founded to permit the setting up of a European telegraph system. Formal standards recommendations have been published since then, covering all fields of telecommunications. Whilst its primary role is establishing international standards in telecommunications and the dissemination of technical information, it is increasingly concerned with promoting international exhibitions having a high technical content. Since 1900 it has contributed to international world fairs and expositions, including Paris 1900, Brussels 1958, Osaka 1970, and a number of smaller national exhibitions in Germany, France and Switzerland. Since 1971, however, it has held its own World Telecom Exhibitions

---

[2]International Telecommunications Union, Place des Nations, CH-1211, Geneva 20, Switzerland

every four years to coincide with the plenary sessions of the standards-making body, the CCITT. These are large technical and educational exhibitions open to the public and commanding a premier place in the modern world of communications and information technology. The ITU Telecom Exhibition attracts some 200 000 participants and is now the world showcase for telecommunications innovators, with the equipment demonstrated often resulting from the realisation of the work of the standards committees held in the previous meeting.

The first of these, Telecom'71, was held between 17 and 27 June 1971 in the Palais des Expositions in Geneva, occupying some 24 000 m^2 of exhibition space with 250 exhibitors. Much of the educational material was shown at the opening to a worldwide television audience of an estimated 400 million viewers and repeated in the colour television studios throughout the exhibition. Satellite transmission was utilised to demonstrate videophones by companies from Germany, France, Sweden and the United States. Multiplex television was shown by the Japan Broadcasting Corporation and was a new phenomenon of the use of waveguides for television transmission—now superseded by optical fibres—and in the 1970s was considered an exciting wide-bandwidth method of data transmission. As early as 1971 the theme of exhibiting for the 21st century was taken up by the individual exhibitors, who vied with each other to show the telecommunications research that had been carried out in member countries [12]. The exhibition covered many fields relating to telecommunications, radio and television broadcasting, electronics, data transmission, telephone networks, audio-visual media transmission, etc., and was the first exhibition to do so on such a wide international scale.

Subsequent exhibitions in 1975, 1979, 1983, 1987, 1991 and 1995 continued the theme of universal coverage with the emphasis on what is possible for the next exhibition as well as demonstrating the technology currently available. The existing system of worldwide telephony and data transmission was not neglected, and with each exhibition details of the improvements in the infrastructure of the network system were shown by most companies exhibiting, in line with the CCITT recommendations currently agreed. At the same time, elsewhere in the exhibition buildings the ITU committees were meeting to continue discussions on changes in standards required in the next few years to meet the new telecommunications possibilities.

By 1983 this infrastructure had been updated to provide a universal system of pulse-coded modulation of group telephony transmissions;

solid-state exchanges were in place using a new system of stored program computers, and the Telecom Exhibition had moved to a new $73\ 000\ \text{m}^2$ exhibition hall, Palexpo in Geneva. In 1991 further development in global networking was leading to global services interconnectivity, with optical tranmission systems, satellite techniques, cellular radio, and personal communications, all of which were to be demonstrated on the stands of this new enlarged exhibition [13].

A prime concern with Telecom Exhibitions in recent years has been software and its implementation in various communication networks —private, national and international. At Telecom'88 the emerging Open Systems Interconnection (OSI) systems were being supported by British Telecom and other agencies with office communications systems such as MEZZA, which fully integrates voice, data communications and computing in the 'electronic office'—a term which was beginning to be implemented in many business concerns. This exhibition also saw the emergence of FAXMAIL, with demonstrations provided by British Telecom.

The recent Telecom'95 at Geneva featured interconnection with the worldwide Internet and wide-band Integrated Systems Digital Networks (ISDN). The facilities provided by communications software are as interesting to visitors to this trade fair as the hardware exhibited by many different manufacturers (Figure 14.4). New features were ORSPEED, a system which provides a data rate of up to 52 Mbit/s over ordinary copper wires for short distances. This is a much needed facility to enable connection to wideband fibre-optic networks (now also available in many developing countries), which for practical reasons are not yet able to be connected directly into a large number of homes. Videophone systems were shown by Nokia as one of many new developments which are able to combine several low-speed data channels into a wideband transmission network. Transmission bandwidth and its availability was a key factor in many of the new devices on show and is obtained in a variety of different ways. One of these is satellite telemetry, and an imaginative application was shown by GPT Telemetry Systems to link together vending machines, now an indispensable part of many retail establishments and shops (e.g. Coca-Cola machines), so that data from these can be monitored from a central facility to carry out stock control. A new feature shown at Telecom'95 was the possibility of visiting an exhibition on an interactive basis through the use of a teleconferencing facility. About 80 UK firms took part in this development, which enabled exhibition details to be transmitted over a computer network from Geneva for display on a computer screen, and enabled exchange of documents to

INTERNET AT TELECOM 95
7 & 8 OCTOBER, GENEVA–SWITZERLAND
ORGANIZED BY THE INTERNATIONAL TELECOMMUNICATION UNION

Key Speakers include:

Barry Berkov, Executive VP, CompuServe • **Brian Carpenter,** Chairman, Internet Architecture Board
David Chaum, CEO, DigiCash • **Jim Clark,** CEO, Netscape • **Elon Ganor,** CEO, VocalTec
Christian Huitema, W3O Consortium • **Scott Kurnit,** Senior VP, MCI Information Services
Bruno Lanvin, World Coordinator, World Trade Points Programme, UNCTAD • **Tony Rutkowski,** Executive Director, Internet Society
Lee Stein, CEO, First Virtual • **Wim Vink,** Managing Director, EUnet Communications Services.

Also Representatives from: Europe Online, Apple eWorld.

And presentations on:

Sun Microsystems – **Hot Java** • VocalTec – **InternetPhone**
Silicon Graphics – **Virtual Reality on the World Wide Web**

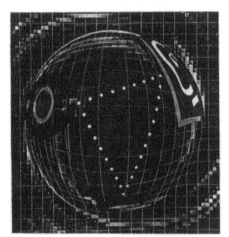

You've been promised the world before. Now it's really yours. Right on your computer screen. With the Internet.

Internet @ TELECOM 95 will show you this new world already shared by over 20 million Internet users. It's a special weekend session of the TELECOM 95 Forum which together with the Exhibition runs from 3-11 October.

You'll hear the leading Internet experts discuss the evolution and growth potential of the Internet, the strategies of the big carriers and On-line service providers. You'll also learn about the commercial and regulatory issues. And you'll discover from the top industry specialists the new wave of Internet applications that are driving public and commercial interest. What's more, you'll get hands-on experience at the Internet Cafe.

We've also made it easy for you to gain full access to the vast TELECOM 95 exhibition during the weekend of the Internet Forum. A two-day pass is included in the Forum fee. Here you'll have a privileged perspective of the technologies driving the entire telecommunications industry. It's also a rare opportunity to visit the hundreds of exhibitors of Internet services and applications at TELECOM 95.

The weekend of October 7 and 8 is a moment to seize. To register, simply fill out the coupon and mail it to the address below or fax it to +41 22 730 5926 or register directly on the WWW address URL:http://www.itu.ch/TELECOM.

For full program details, dial the Fax on Demand No. +41 22 730 6464 from the handset of a fax machine or phone +41 22 730 6161 or connect to the Telecom World Wide Web address above. **MEET THE NEW PLANET EARTH.**

Figure 14.4   Telecom'95 at Geneva
*(Br. Telecom J., 1995, **16**)*

take place over an associated FAX system, which included form-filling and placing of orders [14].

14.7 The Hannover Messe

A highly specialised and yet extensive exhibition, typical of the trade fairs held in the 1990s, is the Hannover Messe, where over 6000 exhibitors from 50 countries bring together just 12 related areas of industrial technology.

The rise of this huge trade fair has been quite remarkable. The first Hannover Messe was held in 1947 by making use of some disused buildings in the bombed city with the intention of encouraging

German industry at the end of the War to restart its activities in peaceful production. The Hannover Messe is now, in effect, a large collection of specialised fairs generally having one collective theme, a typical example being the exhibition held in 1995, which consisted of eight specialised exhibitions:

1. electric automation technology; 1250 exhibitors from 35 countries and requiring four halls
2. power transmission and control; 1200 exhibitors from 29 countries (eight halls)
3. energy-environmental techniques; 475 exhibitors from 22 countries (three halls)
4. plant engineering; 365 exhibitors from 16 countries on an open air site
5. factory equipment and tools; 460 exhibitors from 26 countries (two halls)
6. subcontracting and industrial materials; 1550 exhibitors from 45 countries (three halls)
7. lighting technology; 650 exhibitors from 28 countries (three halls)
8. research and technology; 550 exhibitors from 21 countries contained in a single hall.

Considerable reliance is placed on the use of interactive information systems and computers to assist the business visitor. These are arranged to produce a tour guide based on information entered by the visitor and can be prepared as a product or exhibitor search and arranged before the visitor reaches the location by making use of an information disc sent by post[3].

This facility is also a feature of another modern Hannover trade fair, Ceebit, where 5700 companies from 52 countries exhibit only computers, software and information technology. It provides, year-by-year, striking examples of innovation, in most cases accompanied by demonstrations which have a high educative content. An example from the 1995 Ceebit fair is provided by Siemens and seeks audience participation in a new development of three-dimensional graphics displayed on a computer screen. This makes use of a sensor above the screen directed so as to determine the orientation of a viewer's head. An image of a three-dimensional object displayed on the screen then rotates as the viewer's head moves to one side or the other, thus providing a convincing three-dimensional effect.

---

[3]As with several of the large modern trade fairs information can also be obtained in advance by access to the Internet. For the Hanover Messe information on the exhibitors and products can be accessed under http://www.mess.de

14.8 Düsseldorf INTERKAMA

Düsseldorf has been an important trade fair centre in Germany since the middle ages. Its new Trade Fair Centre is the venue for more than 40 trade fairs each year, with the INTERKAMA Exhibition of Measurements and Automation as one of the largest. The first INTERKAMA exhibition was held in 1957 as an International Congress with an Exhibition for Measurement and Automation, initiated by the Düsseldorfer Messegesellschaft, Siemens AG and Bayer AG, and attracted over 41 000 visitors. It has been held at three-year intervals since then, with increasing attendance and participation from many countries. The international role of the exhibition increased after the formation of a European Committee in Measurement and Automation (CEMA) to co-ordinate similar exhibitions from neighbouring countries, with INTERKAMA now the largest trade fair in this specialised area in Europe.

The 1995 Exhibition was contained in 15 major halls and a number of smaller locations, exhibiting equipment for 1250 exhibitors for the five days of its opening, with an attendance of almost 100 000. The International Congress no longer forms part of the INTERKAMA proceedings, its place taken by a series of technical conferences in current procedures and equipment for measurement and automation. More specialised than the Hannover Messe, the exhibits cover a diverse field of interest: control and monitoring systems, sensors, measuring and analysis instrumentation, drives and actuators, open and closed loop control systems, data processing, communication, software and services. The 1995 Exhibition introduced new themes of microelectronics, neural networks and fuzzy logic.

14.9 References

1 FISCHER, M.E.: 'Breitbundkabel mit neuartiger isolation', *ETZ*, 1935, **56**, pp. 1245–1248
2 'The electrical exhibition at Olympia', *The Electr. Times*, 21 Sept. 1911, **40**, pp. 252–293
3 WOODHAM, H.S.: 'Britain's shop window—the BIF' (Newservice, London, 1948)
4 'Exhibitions and fairs', *The Engineer*, 12 Apr. 1946, **181**, pp. 344–345
5 'Component for automation', *Engineering*, 17 May 1957, **183**, pp. 636–638
6 AUGER, H.A.: 'Trade fairs and exhibitions' (Business Publications, London, 1967)
7 News item 'The British Industries Fair', *Engineering*, 1958, **185**, p. 175
8 CATTERALL, P.: 'History of Electrex' (ASEE Exhibitions Ltd., West Horsley, Surrey, 1993)
9 'Sixth Electrical Engineering Exhibition', *Engineering*, 1957, **183**, pp. 470–473
10 See for example 'The Farnborough Air Show', *Wireless World*, 1946, **52**, p. 366

11 SARGROVE, J.A.: 'New methods of radio production', *J. Brit. IRE*, 1947, **7**, pp. 2–33
12 'Message to the XXIst century: Telecom 71', *Telecommun. J.*, 1971, **38**, pp. 683–697
13 'Telecom 91', *Telecommun. J.*, 1991, **58**, pp 503, 613, 683
14 'Computers and telecoms coverage in Geneva for Telecom'95', *IEE Rev.*, 1995, **41** (6), p. 224

Chapter 15
Epilogue

15.1 A summary

In this book we have traced the technical features of exhibitions from their beginnings in medieval times as essentially trading and barter events, often allied with religious festivals, to the present day with its all-embracing technical exhibitions, highly dependent on electronics and information technology. We have looked particularly at the way electrical technology has been shown and considered, initially as a mysterious force to the general public and imperfectly understood by its early practitioners, to its exploitation at the present time for all aspects of the presentation and content of world fairs.

On the way we can recognise significant milestones: the Royal Society meetings and technical gallery exhibitions held in London in the first half of the nineteenth century; the Great Exhibition of 1851; the invention of the Gramme dynamo, enabling brilliant lighting effects to be achieved; specialist electrical exhibitions showing a wealth of application for industry; the public availability of an electrical supply, leading to a widespread use of domestic appliances; the divergence of exhibitions into world fairs, with their sophisticated education and entertainment features; and trade fairs presenting a massive and international 'shop window' to modern electrical development.

In the early Chapters we looked at the representation of electrical technology in individual exhibits or specific parts of the exhibition, often confined to a special electrical section where the generation and application were considered together. Indeed this was an essential feature of large exhibitions following the development of the Gramme dynamo and before the resources of central generating stations became available for the exhibition site. Now with the modern exhibition we find that electricity plays a major part throughout each and every exhibit so that it is no longer practicable or profitable to set out to describe its influence in one particular area. Instead the vast riches of electrical applications technology, electronics, the computer, networks and complex software are considered not only as component parts of artifacts to be displayed, but also as tools to present them in the most effective manner, to operate them and to entertain the exhibition visitor.

15.2 Influence of information technology

To complete this outline of the technical features of exhibitions we need now to consider briefly the direction in which present developments are taking them to obtain some idea of the part that electrical science could be expected to play in the next decade or so.

Chief amongst the influences for change is information technology. Already we are beginning to see such change involving interactive computer enquiry systems, CD-ROM and network-distributed cataloguing, information availability on the Internet, satellite communication and interactive participation, including the use of virtual reality displays, which determine how an individual visitor may be able to form an intensely private view of such 'public' exhibitions. It may well turn out that future exhibitions will become yet bigger but, by means of 'user selection' in this way, that they will appear to satisfy every exhibition visitor regardless of diverse requirements of individual needs, particularly in the area of highly technical trade fairs.

15.3 Computer networking—the Internet

The most significant recent event for international exhibition development lies in the use made of the worldwide computer network, the Internet. Already we have seen its use as an information medium for proposed and running exhibitions (Hannover Messe'96, Lisbon Expo'96, Hannover Expo'2000) and, associated with a facsimile system, for business transactions (Geneva Telecom'96). A more far-reaching idea, however, is to use the Internet as a development 'site' for the exhibition itself. Here a number of organisations or countries each prepare an electronic version of their exhibits, using both still and moving images with explanatory text and commentary, and store these in a digital database. This amalgam can form the exhibition itself, which can then be accessed by the exhibition visitor through a computer workstation linked to the Internet, and thus secure a virtual 'visit' to the exhibition.

This has been carried out recently through two attempts to secure an 'online' exposition via the Internet [1]. The first of these was initiated as an online exposition by Worlds Inc. of San Francisco in 1995 with the not unexpected theme of 'connecting people to people' and described as an 'interactive world's fair':

> *'It offers computer access to a world of virtual pavilions and interactive exhibits. Participants will encounter a navigable 3-D computer graphics landscape of art, technology and cultural pavilions... and includes Biospace, Cyberspace and Timespace pavilions.'*

It is intended to add other nonscientific areas and evolve continuously with new 'virtual pavilions' and changing exhibits as the technique develops.

A second and more extensive event is the Internet 1996 World Exposition, organised by the Internet Multicasting Service of Washington. This promises to become a seminal event, creating an infrastructure, together with networking hardware, which will remain in operation after the exposition ends in much the same way as did several of the larger world fairs since 1851, with their legacy of permanent buildings and structures.

15.4 Internet 1996 World Exposition

The exposition's theme is *'creating a public park for a global village'*, and was conceived by Carl Malamud and Vinton Cerf, an Internet designer, in 1991 [2]. To create this a number of hardware manufacturers, including Sun Systems, Quantum Corporation, Sony, NBC, NTT, NiftyServe, Softbank and others have donated equipment which is situated in various locations: in the Internet Multicasting Service, John F. Kennedy Centre, MIT Media Laboratory, Lincoln Centre, Smithsonian Institute, Tokyo Aquarium and a number of Universities in Japan, the United States and in London. The most significant of these contributions are the provision of 1 Tbyte (10^{12}

bytes) of dynamic data storage by Quantum Corporation and massive server computers by Sun Microsystems. This equipment, forming the 1996 World Exposition, is referred to collectively as 'Central Park', although in practice it is distributed over many sites throughout the world and linked together through the Internet. To facilitate this interconnection a dedicated series of high-speed telecommunications links supplements the general-purpose Internet infrastructure. These are contributed by a number of organisations and use T3 (45 Mbit/s) and other high-speed links currently linking sites in mainland United States, Japan and Europe and across the Pacific and Atlantic, including the use of some satellite links by JapanSat covering the Pacific rim countries [3].

As at previous world fairs, the Internet Expo will feature 'pavilions' containing exhibits on particular themes[1]. Over 100 countries are contributing pavilions together with a number of organisations [4]. An example is the 'global schoolhouse pavilion', which allows children around the world to communicate with each other, commencing with a series of Internet video conferences on the environment between children in England, Virginia and California and including competitions and interactive contributions from schools in many countries. In the 'town hall pavilion' an interactive event will be held on the 50th anniversary of the computer, and the 'world expositions pavilion' will provide information on world fairs of the last hundred years and that proposed at Expo'2000 at Hannover. In a 'reinventing government pavilion', United States government databases on patent, trademark and financial information will be available. Other pavilions include a 'small business pavilion', an 'environmental pavilion' and a number of cultural pavilions contributed by many countries, a notable example being an extensive set of demonstrations of Thai cooking in their 'Aw Taw Kaw Pavilion' [5].

15.5 Exhibition planning for the future

The heady ideas discussed above, many of which have been effectively applied to the expositions and fairs of the 1990s, are of course, limited by financial and political problems which can be, and often are, dominant. This is especially true of conventional major exhibitions. The cost of mounting a world fair is now high, and fewer applications for major events are received by the offices of the BIE each year. Some

---

[1]Internet address: http://park.org/

are planned in detail and then abandoned when the costs are fully realised, examples being the Budapest Exposition, planned for 1996, and cancelled several years earlier, and the Paris Exposition for 1989, which should have been held to celebrate the bicentenary of the French Revolution. The 1889 Exposition was spectacular with its legacy of the Eiffel Tower and acting as a showcase for the burgeoning electrical industry. The Paris 1989 Exposition, with its theme of 'the path of liberty', was agreed with the BIE, despite an application from the United States for its Columbian Exposition to be held in 1992 simultaneously in Chicago and Seville (BIE rules preclude two major world exhibitions being held within five years of one another). It was to be spread out over almost all the sites used by previous Paris expositions from 1855 onwards together with two new sites in Paris, each of about 60 ha, and was expected to attract 62 million visitors. The political will was not, however, found to proceed with the project, and it was cancelled by President Mitterrand in 1983 on financial grounds [6]. Cancellation of an already planned world fair is itself symptomatic of these volatile times. A recent traumatic event (for the organisers) was the cancellation of Tokyo's Expo'96, already under construction, which is likely to generate a compensation bill of several billion yen—a sizable fraction of the cost of holding the actual event as planned [7]. As an insurance against failure of this kind exposition organisers are looking for benefits other than revenue from fickle attendance figures. Seville, in planning its Expo'92, considered the benefits to the region from improved transport facilities and permanent buildings left behind after the exhibition [8, 9]. Similar considerations applied to Expo'85 in Tsukuba, and will be important at the two BIE-registered events, already agreed as Expo'98 in Lisbon and Expo'2000 at Hannover, both of which are described briefly below.

Despite the spate of cancellations a number of recent proposals are going ahead. Some have already received full planning permission from the BIE as registered exhibitions; others are still in the early planning stage.

Stockholm 1997 is being planned on a strong technical theme, 'creative man', and expects to be able to create a new Science Centre for the city, adjoining the existing Technical and Tele Museum, located within the Exhibition grounds, and which will remain after the Exhibition closes.

Lisbon World Expo'98 has received BIE approval and building work has commenced on a 360 ha site outside Lisbon. The central theme will be 'the oceans, a heritage for the future', and will celebrate the voyages of Vasco da Gama, who sailed from Lisbon in 1498[2]. The exposition is endorsed by the President of the United Nations as forming part of the UN's 50th anniversary celebrations. As with all the post-1995 exhibitions the emphasis is on the environment and, in this case, the management of water resources and preservation of the ecological balance. Seven new large pavilion buildings are being constructed for Britain, France, Spain, Greece, the Netherlands and the European Union. This last has its own theme of 'energy comfort 2000', and is intended to show how building design can reduce the need for air conditioning by 50 % and carbon dioxide emission by 50–70 %. A shared 'pavilion of the future' will house an interactive and multimedia theatre to promote the Exposition's theme. Most of the buildings are expected to remain after the exhibition closes and contribute to a planned expansion of the city. In addition to the district heating and cooling, remotely controlled from a central control station, all the buildings will be equipped with a fibre-optic telecommunications infrastructure. This will be the first BIE exhibition to exploit the use of the Internet as an information source before and during the exhibition (Figure 15.1). An attendance of 18 million visitors is expected, and support given from over 30 countries and international organisations, the European Union, United Nations and UNESCO. Again the emphasis is on applications of technology, rather than its demonstration, which is a common feature of all the exposition proposals currently being considered.

15.6 Beyond 2000: exhibitions for the 21st century

A number of exhibitions are planned for the beginning of the next century. A major exhibition at Hannover has been approved by the BIE, and other proposals have been lodged with the BIE from Aichi, Toronto, Calgary and Venice, whilst smaller exhibitions are planned for Denver and the United Kingdom.

Hannover, Expo'2000 will carry the theme 'man-nature-technology', looking at what it regards as the central issues confronting mankind in the 21st century. The Hannover Exposition is fortunate in being able to make use of a well developed site with existing exhibition buildings used for the Hannover Messe (see previous Chapter), together with an infrastructure already capable of accepting up to 200 000 visitors per day. Forty million visitors are expected. To

---

[2]Internet address: http://www.expo98.pt

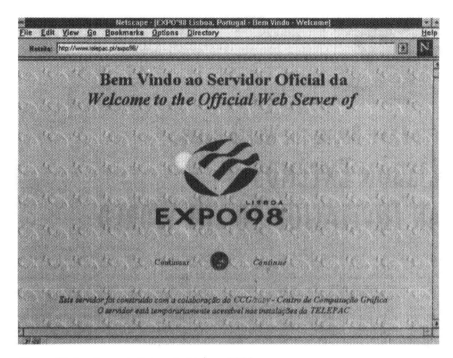

Figure 15.1   *Access to Internet, Lisbon, 1998*
            (Screen photograph)

augment the accommodation available within the city a model village is planned to be built and later to form a new suburb of Hannover. This Expo village of 2500 units is intended to show a model of urban development in all its ecological, economic, cultural and social aspects—a vastly expanded version of Prince Albert's 'model dwellings' built for the Great Exhibition of 1851 in London.

The organisers already have firm ideas on the implementation of the theme and have stated that *'Competitiveness and showy displays of national progress are to be replaced by demonstrations of responsibility, human understanding, the pursuit of worldwide cooperation'.* Modern industrial and scientific developments will, however, be playing a different role and are likely to be incorporated into such ideas as simulated worlds made possible by virtual reality, laser projections and a variety of interactive exhibits. Interactivity is expected to be the leitmotiv of these displays, with *'static exhibits of the conventional type didactically out of place in the Expo theme'* [10].

Some of the world's current projects are expected to be incorporated into suitable exhibitions. Examples already suggested are California's water conservation programme, Zimbabwe's plans to produce bioenergy from sugarcane and Indonesia's search for new ways of protecting the rain forests [11]. The organisers state that

> *'this will be no World Expo of glittering machines and new technical superlatives', but an opportunity to present projects, 'which present innovative, forward-looking answers to the challenges of our time in different regions and cultures of the world'.*

The Denver Expo'2000 expects an attendance of 16 million visitors to a technical exhibition having strong commercial roots in high technology electronics and computers. It is also planning for extensive gardens, an orchestra amphitheatre and a botanic hall which will all remain as legacies after the event.

The proposed Aichi World Expo'2005 at Chubu, 20 km east of Nagoya, also has technology as a prime motive, with an extensive theme of 'technology, culture and communication for the creation of a new global community'. The 250 ha site and its surrounding 2000 ha is being developed into a model city for the 21st century with an initial population of 33 000 and accommodation for a million visitors a year. The organisers regard this as an experimental city which, together with ideas for human societies in the 21st century, they hope to demonstrate will be *'in global harmony with the ecosystem and serve as a global house of wisdom'*, offering a forum for the consideration of worldwide issues. These aspirations will be supported by conferences and symposia on such issues as sustainable development, resource recycling and energy efficiency in which technology will play a major role. A new airport, Chubu International Airport, is scheduled for completion in 2005 and will accommodate an influx of some 25 million visitors expected at the Exposition.

Competing with the Aichi proposal for BIE recognition in the year 2005 is an application from the city of Calgary in Canada for an international exposition with an expected attendance of 12 million visitors to a 162 ha site, located at the confluence of the Bow and Elbow rivers. A theme has yet to be adopted but is likely to take the general line, supported by most current proposals, with its ideas of social responsibility, and expects to *'bring the nations of the world together to explore a future of learning, creation and living together'.*

Plans are also under consideration for a Millenium Exhibition for the year 2000 in the United Kingdom, and both the 1851 Great Exhibition and the 1951 Festival of Britain have been suggested as planning models. Current proposals (1996) are to hold this in a presently derelict site of 21 ha by the Thames in Greenwich, London.

15.7 Change in the characteristics of world fairs

The characteristics of world fairs can be seen from these technical opportunities, decisions and accompanying financial pressures to be undergoing significant changes. Whilst originally conceived as a balanced amalgam of education, entertainment and as a showplace for a nation's skill, ingenuity and industry[3], the mixture is now a somewhat uncertain one with the entertainment features dominating. Substantial sections of the exhibition grounds are often used as a fun fair with the idea of popular entertainment permeating throughout the exhibition, as seen in the case of Vancouver in 1986 and Brisbane in 1988. Beusande–Vincent, writing in *'Le livre des expositions universelles'*, looks back to the 1970s and finds that

> '*as the development of electricity makes way for that of electronics: it manages, it astonishes, it amuses*', and that, '*the new technology becomes just a pretext for fun and amusement*' [2].

Rather more scathing are the remarks made by Alfred Heller, who describes the Louisiana World Fair of 1984 as

> '... *a carnival that calls itself a World's Fair*', and goes on to quote Filson, an architect employed in exhibition design, as observing that, '... *World's Fairs are no longer places for technological announcement and they no longer have anything to do with cultural exchange ... (containing) ... non-stop jazz and a gospel tent*' [12].

In mitigation of these views must be mentioned that the difficulties found in presenting today's new technologies are high. There is the new problem of

---

[3]The 1928 International Convention on Exhibitions defined these as 'An exhibition is a display which, whatever its title, has as its principle purpose the education of the public: it may exhibit the means at man's disposal for meeting the needs of civilization, or demonstrate the progress achieved in one or more branches of human endeavour, or show prospects for the future'

'How to get across new technologies, which today are often highly miniaturised solid-state devices without huge gear wheels, flashing lights and noise that made machine-age displays so attractively kinetic and mesmerising. Exposition display is tending more and more to elaborate audio-visuals (tennis court size television, huge curved screens, walls of slide projectors, laser towers etc. [13].

Also we may note that, where central or government funding is reduced or absent, the competing demands of individual sponsors and their exhibition pavilions can conflict to produce the confusion already noted in connection with the New York World Fair of 1964–1965.

However, in the 1990s we are beginning to find that outside North America and Australia the balance of attractions seems to be maintained fairly well, and regard is paid to exhibiting the new possibilities provided by technology in an informed manner. This is particularly the case with the recent large international exhibitions mounted by Far East countries, such as Japan with Tsukuba Expo'85 and the Korean Taejon Expo'93, which retained the capacity to astound the visitor with its mixture of erudition and display. However, even here technology, whilst being applied to a wide variety of exhibits, many of them working or interactive, tends not to be emphasised as a specific feature. In the case of the Tsukuba Expo'85, for example, the Japanese as a deliberate policy did not exploit fully their undoubted ability in computer technology. The organisers considered that technology had its 'down side'—it has some disadvantages (e.g. environmental)—and were content simply to apply this in entertainment devices such as the popular 'walking, talking and writing robot'. This caution in regard to the application of technology finds its expression in a number of recent expositions. For Seville Expo'92 the organisers gave the view that

'In the middle years of this century it seemed that science and technology, upon which the industrial culture was based, had the power to solve the world's problems. Now we are not so sure. We are more aware of the damage it causes than the benefits it brings. The gap between rich and poor has increased, not diminished, and the growth of cycles of production and consumption is threatening the natural environment upon which it ultimately depends'.

The latest exhibitions and those proposed for the late 1990s and early 2000s do, however, seem to have found a common theme in the consideration of technology to solve some of the pressing environmental and cultural problems and to involve the visiting public with these problems through the medium of interactive displays and events. This is supported by the BIE which, in the mid-1980s, concluded that

> '... *international expositions in the next century should function primarily as forums for addressing issues of global consequence.' It urges that, 'particular attention be given to issues that have resulted from rapid industrialisation and human conflict in this century, such as environmental degradation, population growth and food security'* [14].

Certainly the plans presently released for forthcoming world fairs make little mention of the entertainment features to be expected, with the emphasis placed on using technology as a tool applied to the current problems of the day.

A change in trade fairs is also becoming apparent. These are becoming more informative at all levels and are opening their doors to the general public. In broadening its role to include general educational features in technology the trade fair of the 21st century could be well on the way to providing the modern equivalent of the technology-based international exhibition which took place in many countries at the beginning of the 20th century.

In all cases the world fair and the trade fair can be expected to make extensive use of electrical technology in all its forms to highlight and explain its exhibits, as well as in many cases to create them, so that, far from being an isolated item in an exhibition comprised of a miscellany of competing artifacts, electrical technology has now become, in a real sense, the fair itself.

15.8 References

1 RUBIN, J. (Ed.): 'World's Fair parade of events through 2005', *World's Fair*, 1996, **15** (4), pp. 10–14
2 HAFNER, K.: 'The man with ideas', *Newsweek*, 24 July 1995; see also MARKOFF, J.: *New York Times*, 25 Dec. 1995
3 'A World's Fair for the information age', 1996. Internet Multicasting Service (http://park.org/), Washington; see also 'Internet Expo provides boost to Asia's computer networks', *Nature*, 7 Sept. 1995, **377**, p. 7
4 JOHNSTONE, R.: 'Designing the "global village" at World Net Expo', *New Scientist*, 18 March 1995, p. 19
5 'Internet Exposition', Post Database, Bangkok Post Information Technology, 15 March 1995, Thailand
6 BORDAZ, R.: 'Le livre des expositions universelles 1851–1989'. L'UCAD, Paris, 1983

7 DI SIMINE, M.: 'Tokyo Expo cancellation', *World's Fair*, 1995, **15** (4), p. 7
8 RISPA, R.: 'Expo'92 Seville'. Press Release, Madrid, 1993
9 GRIMSHAW, N., and DAVIS, C.: 'British Pavilion, Seville Expo'92', 1993
10 'The World Exposition Hannover, Expo'2000'. Expo'2000 Hannover GmbH, 30510 Hannover, 1996
11 GARDNER, B.: 'Global community for the Year 2000', *World's Fair*, 1993, **13** (3), pp. 23–24
12 HELLER, A.: 'The Wonder-Wall Fair—Louisiana World Expo'84', *World's Fair*, 1984, **4** (3), pp. 1–4
13 ANDERSON, R., and WACHTEL, E.: 'Expo'86 and the World's Fairs'. Seattle, 1986, p. 32
14 BIE Study, 1995

Index

THE LAST CAR FROM THE EXHIBITION.

Printed in the USA
CPSIA information can be obtained
at www.ICGtesting.com
JSHW011508221024
72173JS00005B/1246